217
Advances in Polymer Science

Advances in Polymer Science
Recently Published and Forthcoming Volumes

Self-Assembled Nanomaterials II
Nanotubes
Volume Editor: Shimizu, T.
Vol. 220, 2008

Self-Assembled Nanomaterials I
Nanofibers
Volume Editor: Shimizu, T.
Vol. 219, 2008

Interfacial Processes and Molecular Aggregation of Surfactants
Volume Editor: Narayanan, R.
Vol. 218, 2008

New Frontiers in Polymer Synthesis
Volume Editor: Kobayashi, S.
Vol. 217, 2008

Polymers for Fuel Cells II
Volume Editor: Scherer, G. G.
Vol. 216, 2008

Polymers for Fuel Cells I
Volume Editor: Scherer, G. G.
Vol. 215, 2008

Photoresponsive Polymers II
Volume Editors: Marder, S. R., Lee, K.-S.
Vol. 214, 2008

Photoresponsive Polymers I
Volume Editors: Marder, S. R., Lee, K.-S.
Vol. 213, 2008

Polyfluorenes
Volume Editors: Scherf, U., Neher, D.
Vol. 212, 2008

Chromatography for Sustainable Polymeric Materials
Renewable, Degradable and Recyclable
Volume Editors: Albertsson, A.-C., Hakkarainen, M.
Vol. 211, 2008

Wax Crystal Control · Nanocomposites Stimuli-Responsive Polymers
Vol. 210, 2008

Functional Materials and Biomaterials
Vol. 209, 2007

Phase-Separated Interpenetrating Polymer Networks
Authors: Lipatov, Y. S., Alekseeva, T.
Vol. 208, 2007

Hydrogen Bonded Polymers
Volume Editor: Binder, W.
Vol. 207, 2007

Oligomers · Polymer Composites Molecular Imprinting
Vol. 206, 2007

Polysaccharides II
Volume Editor: Klemm, D.
Vol. 205, 2006

Neodymium Based Ziegler Catalysts – Fundamental Chemistry
Volume Editor: Nuyken, O.
Vol. 204, 2006

Polymers for Regenerative Medicine
Volume Editor: Werner, C.
Vol. 203, 2006

New Frontiers in Polymer Synthesis

Volume Editor: Shiro Kobayashi

With contributions by

N. Ajioka · K. Endo · H. Kaneko · N. Kashiwa · N. Kawahara
T. Matsugi · S. Matsuo · J. Saito · A. Yokoyama · T. Yokozawa

Springer

The series *Advances in Polymer Science* presents critical reviews of the present and future trends in polymer and biopolymer science including chemistry, physical chemistry, physics and material science. It is adressed to all scientists at universities and in industry who wish to keep abreast of advances in the topics covered.
As a rule, contributions are specially commissioned. The editors and publishers will, however, always be pleased to receive suggestions and supplementary information. Papers are accepted for *Advances in Polymer Science* in English.
In references *Advances in Polymer Science* is abbreviated *Adv Polym Sci* and is cited as a journal.

Springer WWW home page: springer.com
Visit the APS content at springerlink.com

ISBN 978-3-540-69807-4 e-ISBN 978-3-540-69808-1
DOI 10.1007/978-3-540-69808-1

Advances in Polymer Science ISSN 0065-3195

Library of Congress Control Number: 2008933505

Cover design: WMXDesign GmbH, Heidelberg
Typesetting and Production: le-tex publishing services oHG, Leipzig

Printed on acid-free paper

9 8 7 6 5 4 3 2 1 0

springer.com

Volume Editor

Prof. Shiro Kobayashi
R & D Center for Bio-based Materials
Kyoto Institute of Technology
Matsugasaki, Sakyo-ku
Kyoto 606-8585, Japan
kobayash@kit.ac.jp

Editorial Board

Advances in Polymer Science Also Available Electronically

For all customers who have a standing order to Advances in Polymer Science, we offer the electronic version via SpringerLink free of charge. Please contact your librarian who can receive a password or free access to the full articles by registering at:

springerlink.com

If you do not have a subscription, you can still view the tables of contents of the volumes and the abstract of each article by going to the SpringerLink Homepage, clicking on "Browse by Online Libraries", then "Chemical Sciences", and finally choose Advances in Polymer Science.

You will find information about the

- Editorial Board
- Aims and Scope
- Instructions for Authors
- Sample Contribution

at springer.com using the search function.

Color figures are published in full color within the electronic version on SpringerLink.

Preface

It is generally accepted that a new material is often developed by finding a new synthesis method of reaction or a new reaction catalyst. Historically, a typical example may be referred to as a Ziegler–Natta catalyst, which has allowed large-scale production of petroleum-based polyolefins since the middle of the 20th century. New polymer synthesis, therefore, will hopefully lead to creation of new polymer materials in the 21st century. This special issue contributed by three groups focuses on recent advances in polymer synthesis methods, which handle the cutting-edge aspects of the advanced technology.

The first article by Yokozawa and coworkers contains an overview of the reaction control in various condensation polymerizations (polycondensations). Advanced technologies enabled the control of stereochemistry (regio-, geometrical-, and enantio-selections), chemoselectivity, chain topology, and stoichiometry of monomers, giving a high molecular weight polymer. It has been recognized for a long time, however, that polycondensation is a difficult process in controlling the reaction pathway, because the reaction is of step-growth and the reactivity of monomers, oligomers, and polymers are almost the same during the reaction and hence, the molecular weight of polymers and its distribution (M_w/M_n) are impossible to regulate. The authors' group developed a new reaction system (chain-growth condensation polymerization), changing the nature of polycondensation from step-growth to chain-growth; namely the propagating chain-end is active, allowing for control of the product molecular weight as well as the distribution. With a specific initiator and/or catalyst, the chain-growth condensation polymerization came close in behavior to that of an addition polymerization; a M_w/M_n value being even less than 1.2, like a living system, compared with that of a most probable value 2.0 for conventional polycondensations.

The second article by Kawahara and his coworkers focuses on polyolefin (PO)-based hybrid materials (POH), in view of their synthesis, structures, and properties. POs are currently the most widely and conveniently used polymeric materials as recognized by the production amount of over one hundred million tons annually in the world, due to the cheap price yet good properties. They are basically hydrocarbon polymers, and hence hydrophobic and less polar. These basic properties are to be modified by introducing a polar function for a wider use in practical applications. Preparation of POHs is one of the best ways to provide a variety of desired properties with POs. There are three main synthetic routes to POHs; starting from PO macroinitiators, PO macromonomers, and reactive POs. Polymerization or copolymerization using these macroinitiators and macromonomers are carried out by recently developed methods (living polymerization, ATRP, RAFT, NMP, etc) to produce POHs of block, graft, or branched-type structures. Reactive POs can be coupled with other components to give POHs. In addition, living polymerization of olefins can also be a route for POH synthesis. A wide range of applications of these product POHs are exemplified.

The third article by Endo reviews synthesis and properties of cyclic polymers. Polymer structure is basically divided into two classes; linear and cyclic. Combinations of linear and/or cyclic structures yield a variety of architectures such as branched, graft, block, star, ladder, dendritic, catenane, rotaxane and other complicated structures through covalent and noncovalent bonds. A linear or cyclic structure has been an important, fundamental problem since the early stages of polymer science, and yet it is still new. Fortunately, recent developments of analytical methods enabled the structural elucidation of many of these architectures. The synthesis principle to lead to cyclic polymers can be cited as two main methods. One is the utilization of the ring-chain equilibrium, occurring in many polycondensations and ring-opening polymerizations. The other is the end-to-end cyclization (ring-closure reaction) method from α,ω-difunctional linear precursors, via bimolecular or unimolecular processes. Synthesis of a number of cyclic polymers has been achieved by addition, condensation, oxidation, metathesis, and ring-opening polymerizations as well as by polyaddition. A typical difference in physical property between cyclic polymers and linear polymers is also demonstrated.

We hope the readers will learn something new at the forefront of the polymer synthesis field from the above contributions.

May 2008, Kyoto

Shiro Kobayashi

Contents

Adv Polym Sci (2008) 217: 1–77
DOI 10.1007/12_2007_125

Published online: 22 January 2008

Reaction Control in Condensation Polymerization

Tsutomu Yokozawa (✉) · Naomi Ajioka · Akihiro Yokoyama

Department of Material and Life Chemistry, Kanagawa University,
Rokkakubashi, Kanagawa-ku, 221-8686 Yokohama, Japan
yokozt01@kanagawa-u.ac.jp

Abstract Recent progress in the reaction control of condensation polymerization is described. In the past, the reactions in condensation polymerization had been improved for the purpose of production of high molecular weight polymer with more convenient methods. In recent years, chemoselective, regioselective, and stereoselective condensation polymerizations have been developed. Furthermore, it has been revealed that most condensation polymerizations possess the fundamental tendency to yield cyclic polymers as stable end products when the reactions are optimized for quantitative reaction. When the reactivity of a functional group of a monomer was enhanced after the other functional group of the monomer was reacted, condensation polymerization of AA and BB monomers afforded polymer with high molecular weight, even under nonstoichiometric conditions, and condensation polymerization of AB monomer involved a chain-growth polymerization mechanism to yield polymer with narrow molecular weight distribution.

Keywords Chain-growth polymerization · Condensation polymers · Cyclic polymers · π-Conjugated polymers · Polycondensation

Abbreviations

acac	Acetylacetonate
ATRP	Atom transfer radical polymerization
BINAP	2,2′-Bis(diphenylphosphino)-1,1′-binaphthyl
BOC	*tert*-Butoxycarbonyl
Bu	Butyl
cod	1,5-Cyclooctadiene
DABCO	1,4-Diazabicyclo[2.2.2]octane
dba	Dibenzylideneacetone
DBN	1,5-Diazabicyclo[4.3.0]non-5-ene
DBOP	Diphenyl(2,3-dihydro-2-thioxo-3-benzoxazolyl)phosphonate
DBU	1,8-Diazabicyclo[5.4.0]undec-7-ene
DFT	Density functional theory
DMAc	*N*,*N*-Dimethylacetamide
DMI	*N*,*N*-Dimethylimidazolidinone
DMSO	Dimethyl sulfoxide
DP	Degree of polymerization
DPCB-OMe	1,2-Bis(4-methoxyphenyl)-3,4-bis[(2,4,6-tri-*tert*-butylphenyl)phosphinidene]cyclobutene
dppe	1,2-Bis(diphenylphosphino)ethane
dppf	1,1′-Bis(diphenylphosphino)ferrocene
dppp	1,3-Bis(diphenylphosphino)propane
DSC	Differential scanning calorimetry
EB	Emeraldine base
ee	Enantiomeric excess
EPR	Electron paramagnetic resonance spectroscopy
ES	Emeraldine salt
FT-IR	Fourier-transform infrared
GPC	Gel-permeation chromatography
HH	Head-to-head
HPLC	High-performance liquid chromatography
HT	Head-to-tail
LB	Leucoemeraldine base
LiHMDS	Lithium *N*,*N*,*N*,*N*′,*N*′,*N*′-hexamethyldisilazide
MALDI-TOF	Matrix-assisted laser desorption/ionization time-of-flight
MOM	Methoxymethyl
NMP	*N*-Methylpyrrolidinone

OLED	Organic light-emitting diode
PBT	Poly(butylene terephthalate)
PEG	Poly(ethylene glycol)
PEN	Poly(ethylene naphthalene-2,6-dicarboxylate)
PET	Poly(ethylene terephthalate)
Phbox	2,2′-Isopropylidenebis(4-phenyl-2-oxazoline)
PMA	Poly(methyl acrylate)
PPO	Poly(1,4-phenylene oxide)
PPV	Poly(phenylene vinylene)
PTC	Phase transfer catalyst
Py	Pyridine
SEC	Size-exclusion chromatography
SEM	Scanning electron microscopy
tacn	1,4,7-Triisopropyl-1,4,7-triazacyclononane
TBAF	Tetrabutylammonium fluoride
TFA	Trifluoroacetic acid
TFSA	Trifluoromethanesulfonic acid
TH	Tail-to-head
THF	Tetrahydrofuran
TMEDA	*N*,*N*,*N*′,*N*′-Tetramethylethylenediamine
tpzb	Hydrotris(3,5-diphenyl-1-pyrazolyl)borate
TT	Tail-to-tail
XRD	Powder X-ray diffraction

1 Introduction

In the past, most of the research on condensation polymerization was centered on development of methods for the synthesis of high molecular weight polymers. Therefore, reaction control in condensation polymerization was to pursue highly efficient condensation reactions, because high molecular weight polymer is not obtained unless more than 99% of the degree of reaction is attained without side reactions, according to the basic principle established by Carothers and Flory [1–3]. Recent years have seen developments in chemoselective, regioselective, and stereoselective condensation polymerizations as those selective reactions have been developed in organic chemistry.

Other things that are particularly controlled in condensation polymerization are chain topology and cascade reactivity of monomers. In general condensation polymerization, not only linear polymers but also cyclic polymers can be produced because both ends of the polymer are reactive during polymerization. When cascade reactions selectively occur on a monomer, condensation polymerization proceeds in a way different from general polymerization behavior based on the principle of Carothers and Flory. Thus, in condensation polymerization of AA and BB monomers, if AA monomers successively react with both the functional groups on a BB monomer, a polymer with high molecular weight is obtained even at high feed ratio of BB (nonstoichiometric condensation polymerization). In condensation polymerization of AB monomer, by contrast, when both the functional groups on an AB monomer successively react with the polymer end group and another monomer, this condensation polymerization becomes chain-growth polymerization, such as living polymerization, to give polymer with low polydispersity (Scheme 1). Controlling the reaction in condensation polymerization can change the molecular weight of the polymer, the structure of the repeating units, and even further the polydispersity of the polymer. We describe these reaction controls in condensation polymerization in this order.

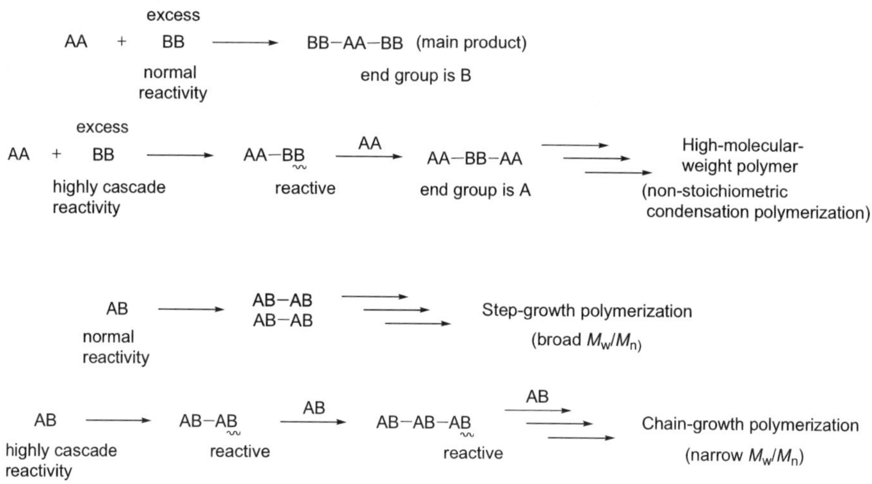

Scheme 1

2 Development of Catalysts

A variety of condensing agents, such as triphenylphosphite/pyridine [4, 5] and DBOP/triethylamine [6], which are used in amounts equivalent to functional groups on monomers, had been developed by the 1980s, whereas new catalysts that promote condensation polymerization exactly with a catalytic amount were developed in the last decade.

2.1 Catalysts for Transesterification and Dehydration

Transesterification was used for many decades for the technical syntheses of PET and PBT. The most widely used catalyst is $Ti(OBu)_4$ (or other tetraalkoxides) alone [7, 8] or in combination with cocatalysts such as La(acac) [9], $Hf(acac)_3$ [9, 10], or NaH_2PO_4 [11]. Kricheldorf et al. studied the influence of various catalysts on the polymer end groups and on the extent of cyclization (also see Sect. 5, Control of Chain Topology) in the condensation polymerization of dimethyl terephthalate with a 300 mol% excess of 1,4-butanediol in bulk up to 250 °C [12]. The highest molecular weights were obtained with $Ti(OBu)_4$, but the MALDI-TOF mass spectra revealed that cycles were the main reaction products below 5000 Da. On the other hand, clean telechelics having two OH end groups were obtained with Bi_2O_3 and bismuth(III) acetate (Scheme 2). The telechelics are useful starting materials for the syntheses of A–B–A triblock copolymers and multiblock copolymers and for a variety of chain extension reactions. With regard to biomedical applications, it is noteworthy that bismuth is seemingly the least toxic heavy metal.

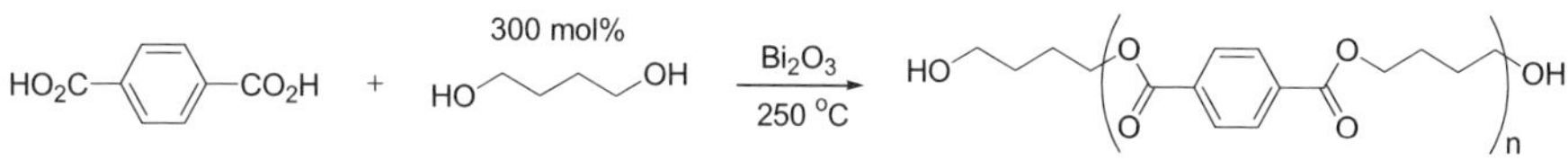

Scheme 2

The most widely used alternative approach to polyester is based on an acid-catalyzed polyesterification of dicarboxylic acids and diols with azeotropic removal of the liberated water. PET has been produced by the transesterfication of dimethyl terephthalate with ethylene glycol. Nowadays, PET is synthesized mainly by the esterification of purified terephthalic acid with ethylene glycol. Antimony in different forms, mainly oxides and acetates, is used to catalyze the condensation polymerization step in most PET plants [13]. Many authors have published results on catalyst activities [13–19]. It is generally accepted that titanium is the most active catalyst, followed by tin, germanium, and antimony. Transesterfication catalysts, such as manganese and zinc compounds, are less active in condensation polymerization environments and strongly boost side reactions. Regarding aliphatic polyesters, which are currently receiving considerable attention due to their biodegradability and the nontoxic character of their degradation products, Otera et al. reported that 1,3-disubstituted tetraalkyldistannoxane **1** [20] was effective for polyesterification of L-lactic acid [21]. The polymerization was carried out in decalin at 190 °C for 24 h to give poly(L-lactic acid) with M_w of 78 000. The high activity of **1** is attributed to no reverse ester cleavage reaction due to its hydrophobicity. Succinic acid and 1,4-butanediol also polymerized with **1** in decalin at 190 °C for 72 h to afford poly(butylene succinate) with M_w of 147 000 [22]. The longer reaction time of this polymerization could be shortened by means of microwave irradiation; a polyester with M_w of 23 500 was obtained at 220 °C for 20 min (Scheme 3) [23].

Ishihara and Yamamoto et al. investigated the catalytic activities of various metal salts for esterification of equimolar aliphatic carboxylic acid and

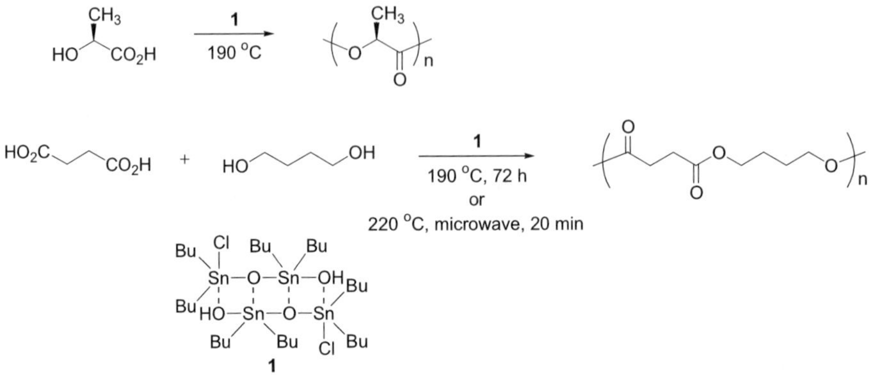

Scheme 3

alcohol in toluene at reflux, and found that $HfCl_4$, $Hf(Ot\text{-}Bu)_4$, and $ZrCl_4$ were highly effective catalysts [24]. Their catalytic activities were higher than those of Ti(VI) and Sn(IV) salts, whereas $HfCl_4$ did not catalyze trans-esterification at all. These results can be accounted for by assuming that the active intermediates that are generated in esterification are Hf(IV) carboxylate and Ti(IV) alkoxide, respectively. Polyesters were synthesized by polycondensation of ω-hydroxycarboxylic acids or equimolar amounts of aliphatic dicarboxylic acids and aliphatic diols in the presence of 0.2 mol % of $HfCl_4\cdot(THF)_2$, which is commercially available and hydrolytically more stable than $HfCl_4$, in *o*-xylene with removal of water for 1 day. The M_w of the obtained polyesters was 34 000–72 400 (Table 1). For polyamides, 3,4,5-

Table 1 Direct polyesterification catalyzed by $HfCl_4\cdot(THF)_2$

Polyester	Isolated yield (%)	M_n^a	M_w^a
$+CO(CH_2)_9O+_n$	95	18 200	34 000
$+CO(CH_2)_{11}O+_n$	97	27 700	72 400
$+CO-(CH_2)_2-CO_2(CH_2)_6O+_n$	98	22 400	38 700
$+CO-(CH_2)_7-CO_2(CH_2)_{10}O+_n$	97	26 900	58 300
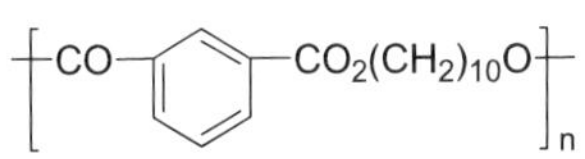	96[b]	13 400	65 100

The polymerization was carried out in the presence of 0.2 mol % of $HfCl_4\cdot(THF)$ for 1 day.
[a] Determined by GPC based on polystyrene standards (eluent: THF)
[b] The reaction was carried out in the presence of 1 mol % of $HfCl_4\cdot(THF)_2$ for 4 days

trifluorophenylboronic acid (**2**) [25] was effective. For example, isophthalic acid and 4,4′-oxydianiline polymerized in the presence of 1 mol% of **2** in a mixed solvent of *m*-terphenyl and *N*-butyl-2-pyrrolidinone at 300 °C under a slow argon flow to remove water for 2 days to yield a high molecular weight aramide ($M_w = 173\,000$) (Scheme 4). When tetracarboxylic acids were used instead of dicarboxylic acid under similar conditions, polyimides were also obtained [26].

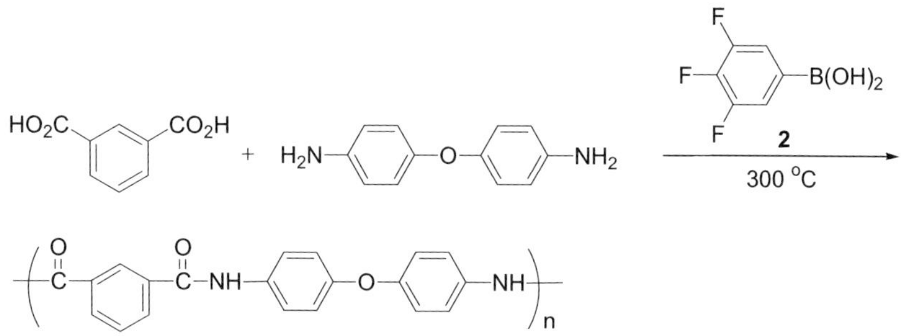

Scheme 4

The above acid-catalyzed polycondensations were carried out at more than 100 °C, whereas Takasu et al. reported room temperature polyesterification with scandium trifluoromethanesulfonate [$Sc(OTf)_3$] or scandium trifluoromethanesulfoimide [$Sc(NTf_2)_3$] [27, 28]. Thus, the direct polycondensation of methylsuccinic acid and 1,4-butanediol proceeded in bulk under reduced pressure (0.3–30 mmHg) using 1.4 mol% of $Sc(OTf)_3$ at 35 °C for 96 h to afford poly(butylene methylsuccinate) with M_n of 12 400 (Scheme 5). When $HfCl_4{\cdot}(THF)_2$ was used in this room temperature polymerization instead of $Sc(OTf)_3$, only low molecular weight polyester ($M_n = 1100$) was afforded. The scandium catalysts did not promote transesterification; ethanol selectively reacted with acetic acid even in the presence of equimolar methyl acetate.

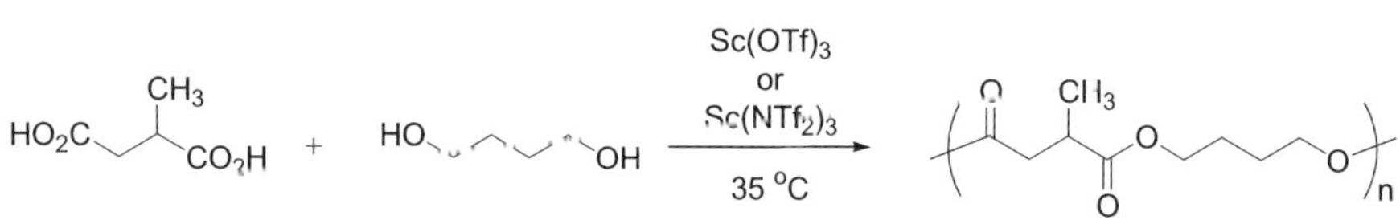

Scheme 5

Friedel–Crafts aromatic substitution reactions have been widely explored in polymer chemistry [29, 30] and generally proceed with Lewis acids such as $AlCl_3$ with elimination of hydrogen halides. In superacid solutions, however, the Friedel–Crafts reactions take place with dehydration from the oxygen of the carbonyl group and the proton of aromatics. The reactivity of the protonated carbonyl group in the superacid can be further increased by the presence of an electron-withdrawing group [31–35], contrary to conventional Lewis acid-catalyzed electrophilic reactions in organic solvents, in which a stable carbocation substituted with an electron-donating group is predominantly generated to react with nucleophiles. For example, 1,2-dicarbonyl groups form highly reactive intermediates in superacidic TFSA [36–39]. Colquhoun and Zolotukhin et al. reported the first application of superelectrophilic chemistry for polymer preparation by reaction of isatin (**3**), containing 1,2-dicarbonyl groups, with aromatic hydrocarbons such as diphenyl ether, biphenyl, or *p*-terphenyl in a TFSA medium at room temperature [40, 41]. The resulting polymers showed high glass transition (T_g) values (337 °C for the polymer from diphenyl ether, and >400 °C for those based on biphenyl and *p*-terphenyl) and the materials obtained were of high molar mass, with inherent viscosities (η_{inh}) in the range 0.31–0.73 dL g^{-1}. Furthermore, ether ketone-type monomers, such as 4,4′-diphenoxybenzophenone (**4**) and 1,4-bis(4-phenoxybenzoyl)benzene, exhibited very high selectivity for linear, all-*para* polycondensation with **3** and afforded poly(arylene oxindole)s of very high inherent viscosity (up to 1.8 dL g^{-1}) (Scheme 6). Smet et al. prepared a monomer consisting of the isatin and phenoxy benzophenone moieties and polymerized it in TFSA to give hyperbranched poly(arylene oxindole)s with a degree of branching of 100% [42].

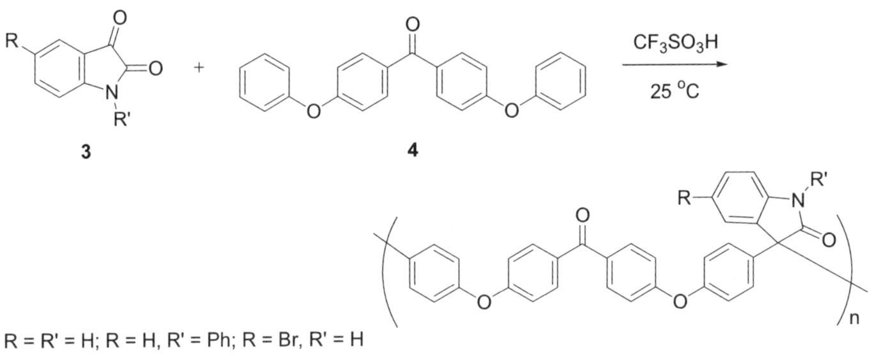

Scheme 6

Cyclic α-diketone, acenaphthenequinone (**5**), also likewise generates highly reactive electrophiles in TFSA and reacts with aromatic hydrocarbons (Scheme 7) [43–45]. However, the reaction of **5** with diphenyl ether or diphenoxybenzene did not afford high molecular weight, linear polymers, whereas the condensation of **5** with 4,4′-diphenoxybenzophenone, a compound which is less nucleophilic than the former aromatics, resulted in linear, high molecular weight, film-forming polymer. In a superacid medium, not only electrophiles such as ketones but also nucleophiles such as aromatic hydrocarbons undergo protonation. Obviously, such protonation of

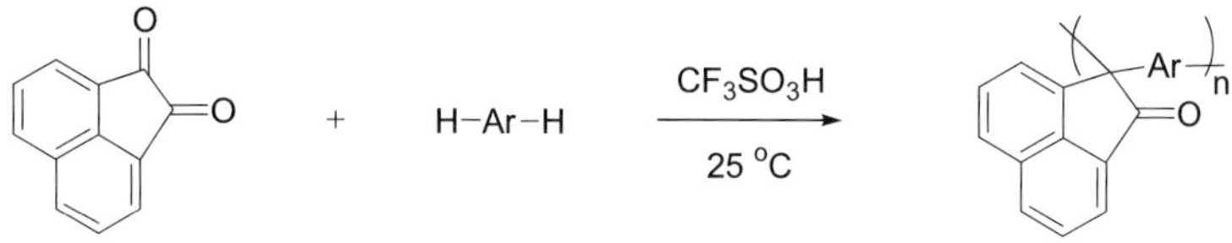

5

Scheme 7

the aromatic would decrease its nucleophilicity and, therefore, prohibit an electrophilic substitution reaction. Thus, in the case of diphenyl ether and diphenoxybenzene, the first protonation occurs at the C4 carbons, while for 4,4′-diphenoxybenzophenone the first protonation site is carbonyl. Therefore, in the case of the former monomers, their protonation impedes polymerization while the latter monomer maintains its polymerizability when protonated. However, a decrease of the acid strength would decrease protonation probability. Indeed, the polycondensation of **5** and activated aromatics, such as diphenyl ether or diphenoxybenzene, carried out in a mixture of TFSA with TFA gave linear, high molecular weight products [45]. The H_0 (Hammett acidity function) of pure TFSA is –14.1, while the acidity of the mixture of TFSA with TFA (44:56, w/w) used was –11.5 [38, 46]. Therefore, reaction media of acid strength H_0 equal to or slightly less than –11.5 are necessary for polycondensations of **5** with activated hydrocarbons. In addition, preparation of polymer from **5** and the less nucleophilic aromatics mentioned above in a mixture of TFSA with TFA was also successful. Acenaphthenequinone-based AB_2 monomers also gave 100% hyperbranched polymers [47]

Carbonyl compounds bearing electron-withdrawing substituents other than 1,2-dicarbonyl compounds were also used for this polymerization chemistry. An equimolar mixture of 2,2,2-trifluoroacetophenone with 4,4′-diphenoxybenzophenone in a TFSA medium at room temperature for 10 h afforded a linear, soluble, film-forming, high molecular weight polymer. The polymer displayed T_g at 175 °C. Under these conditions, 2,2,2-trifluoroacetophenone reacted even with biphenyl to give polymer. Inherent viscosity and T_g were found to be 0.32 dL g^{-1} and 326 °C, respectively (Scheme 8) [48]. Quantum mechanics calculation of model reactions between acetophenone and 2,2,2-trifluoroacetophenone with biphenyl revealed that electron-withdrawing groups favor formation of σ-complexes, which is the limiting step of the electrophilic aromatic substitution reaction. This shows that the adjacent electron-withdrawing trifluoromethyl group enhances significantly the electrophilicity of the carbocation, formed from the carbonyl group, toward nonactivated aromatic hydrocarbons [48–50]. Other fluorinated carbonyl compounds, such as 1,1,1-trifluoroacetone, 1,2,3,4,5-pentafluorobenzaldehyde, and octafluoroacetophenone, were also effective for superacid-catalyzed polycondensation [51]. Electron-withdrawing groups, nonadjacent to the carbocation center, also increase the electrophilic-

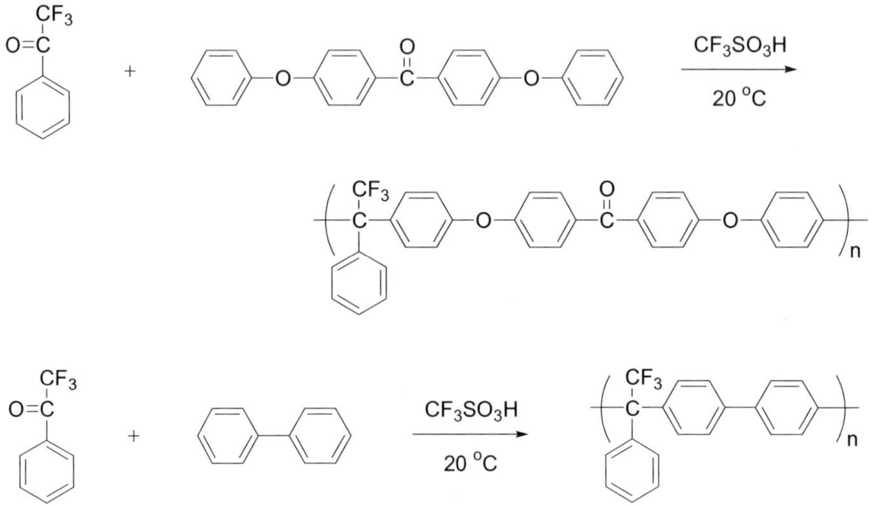

Scheme 8

ity of the carbocation. Thus, 2,7-dinitro-9-fluorenone reacted with 4,4′-diphenoxybenzophenone to produce the high molecular weight polymer with an inherent viscosity of 0.38 dL g^{-1} and T_g of 270 °C (Scheme 9) [48].

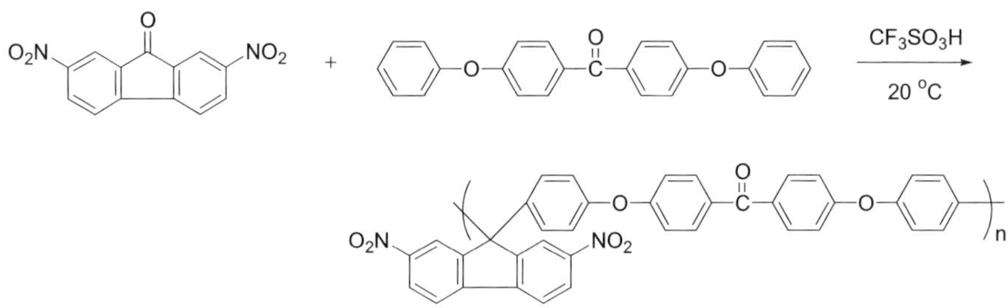

Scheme 9

2.2 Transition-Metal Catalysts

Transition-metal-catalyzed synthesis of poly(arylene)s via carbon–carbon coupling reactions was started by Yamamoto et al. three decades ago [52, 53]; since then various carbon–carbon bond formation processes with transition-metal catalysts have been applied to polycondensation [54–57]. In recent years, Buchwald et al. and Hartwig et al. developed Pd-catalyzed amination and etherification of aromatic halides by using bulky, electron-rich phosphine ligands [58–60], and this chemistry has been applied to polycondensation; for example, poly(triarylamine) as a hole-transporting material in OLEDs was easily synthesized by this method.

Kanbara et al. first reported the polycondensation of dibromobenzenes and aliphatic secondary diamines with $PdCl_2[P(o\text{-tolyl})_3]_2$, but the M_w of the polymer was not higher than 10 000 [61]. When the catalyst was changed to a zero-valent palladium complex, such as a combination of $Pd_2(dba)_3$ and $P(o\text{-tolyl})_3$ or $P(t\text{-butyl})_3$, the yield and molecular weight of polymer were significantly improved; the M_w reached 14 100 (Scheme 10). In the polymerization with the Pd(II) catalyst, the formation of arene from aryl bromide during reduction of the Pd(II) precursor to Pd(0) occurred [62]. The zero-valent palladium complex suppressed this undesirable reaction [63]. The polycondensation using aromatic secondary diamines instead of aliphatic diamines was conducted by Goodson and Hartwig to synthesize polymeric triarylamines that display important conductive and magnetic properties [64]. The polymerization was carried out in the presence of 2% of $Pd[P(o\text{-tolyl})_3]_2$, 2% of $P(o\text{-tolyl})_3$, and 1.5 equivalents of *tert*-BuONa. When *m*-diaminobenzene monomers were used, cyclic oligomers were obtained in the polymerization with both *m*- and *p*-dibromobenzene. On the other hand, *p*-diaminobenzene monomers appeared to be less susceptible to the cyclization than the *meta* analogues; the highest M_w was 45 400 (Scheme 11). However, all of the polymers showed some contamination of the linear material with cyclic oligomers, as evidenced by the multimodal GPC chromatogram. Polymers obtained from bis(amine) monomers and 4,4′-dibromobenzophenone are all bright yellow or orange and exhibit strong fluorescence, making them possible candidates for emissive components in electroluminescence devices.

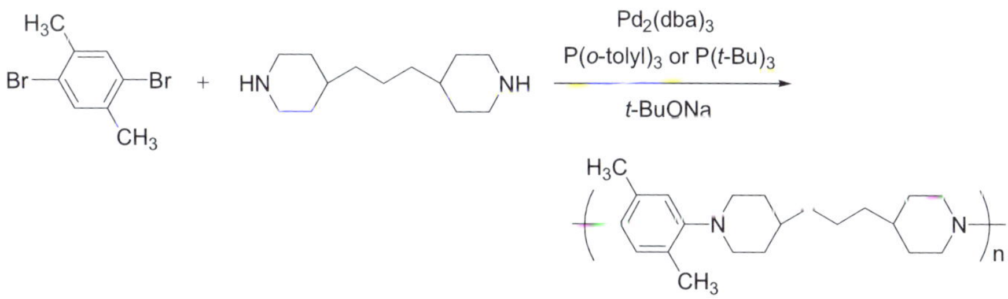

Scheme 10

Polycondensation of aromatic primary diamines and aromatic dibromides gave high molecular weight polyamine ($M_w > 15\,000$) irrespective of the *m*- and *p*-substituted positions of both monomers (Scheme 12). In this polycondensation, BINAP was effective as a ligand of the Pd(0) catalyst; use of $P(t\text{-butyl})_3$, which was effective for the above polymerization using secondary diamines, afforded polymer in a low yield [65, 66]. When BINAP was used

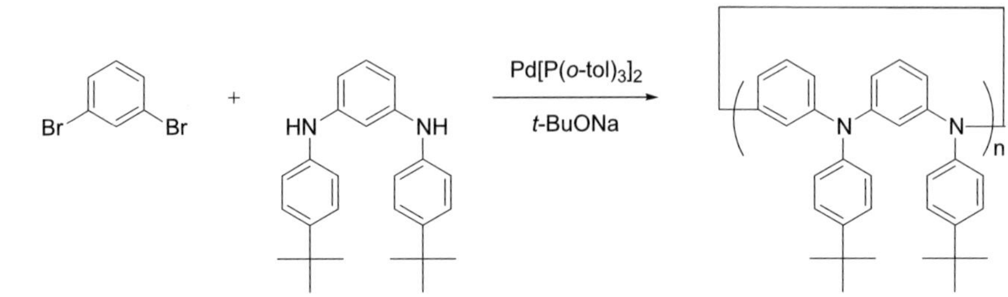

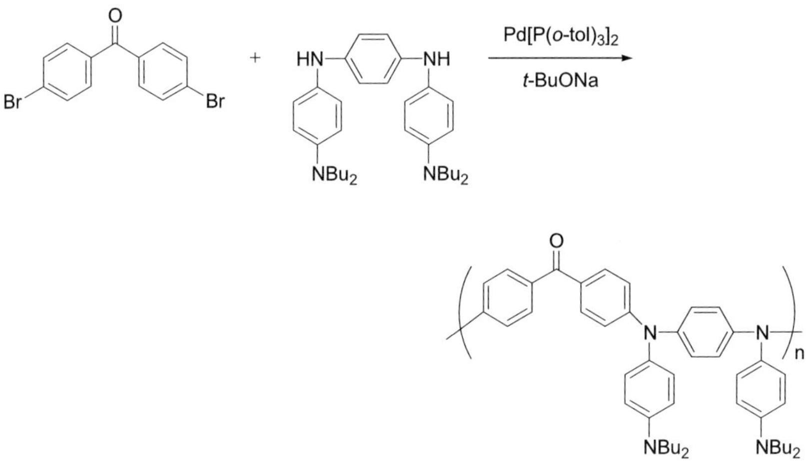

Scheme 11

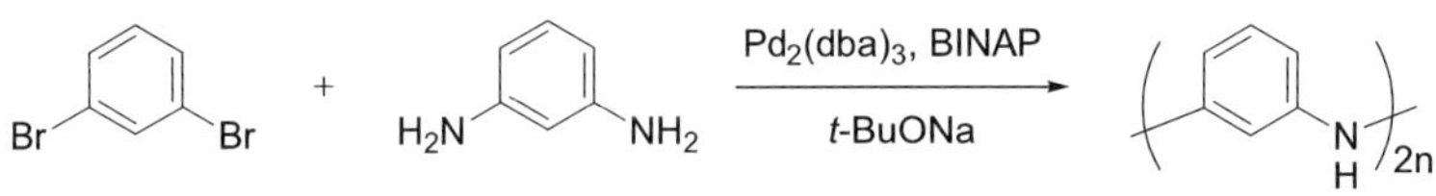

Scheme 12

with $Ni(cod)_2$ instead of $Pd_2(dba)_3$, aryl dichlorides were able to react with primary diamines [67].

Photocatalytically active polymer having acridine units in the backbone was synthesized by the polycondensation of 3,6-diaminoacridine (proflavine) as a diamine and *m*-dibromobenzene (Scheme 13) [68]. Müllen and coworkers synthesized a series of novel poly(imino ketone)s via Pd-catalyzed polycondensation of aromatic dichloro or dibromo ketones with various aromatic diamines (Scheme 14) [69]. The M_w values were in the range of 19 500–474 500. The FT-IR spectra of the obtained polymer revealed that in the solid state intermolecular and intramolecular hydrogen bonding (N–H···O=C)

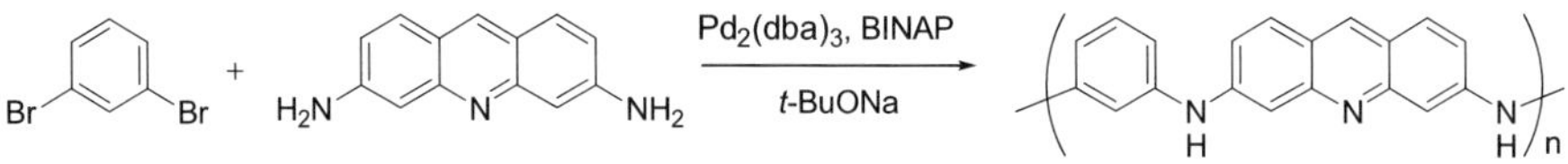

Scheme 13

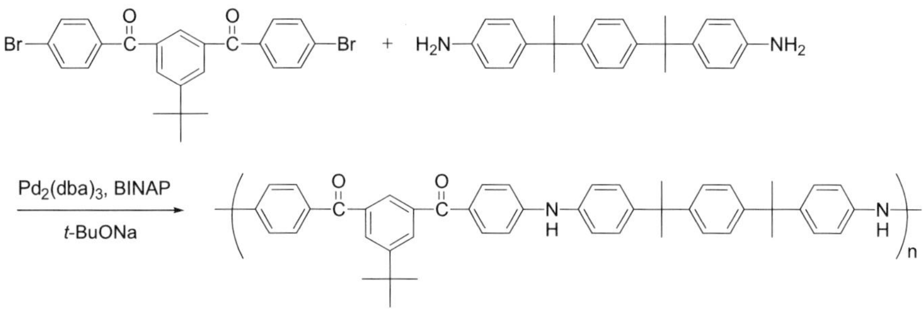

Scheme 14

are present, and therefore the poly(imino ketone)s can be regarded as an analogue to polyaramides. Actually, the polymers were thermally stable up to 323–494 °C and showed high T_g ranging from 202 to 227 °C. The solubility, however, was enhanced. Ward and Meyer reported that a urea-protected trimeric aniline monomer **6** with *para–ortho–para* connectivity and bromine end groups was condensed using a Pd catalyst with a di-BOC-protected trimeric aniline comonomer **7** with all-*para* connectivity and amine end groups to give $(para)_5$-*ortho*-polyaniline with the protecting groups (Scheme 15) [70]. The BOC groups were removed by heating at 185 °C. Analogous to polyaniline, the 5-*p*,*o*-polyaniline exhibited three distinct oxidation states. This result suggests that the presence of the *ortho*-substituted aniline ring does not critically disrupt the conjugation of the polymer or interfere significantly with interchain interactions.

Aromatic primary monoamine such as aniline serves as a difunctional monomer for Pd-catalyzed amination polycondensation to afford poly(triarylamine)s. For example, *m*-dibromobenzene [71] or 2,7-dibromo fluorenes [72] reacted with aniline derivatives (Scheme 16). In these polymerizations, P(*t*-butyl)$_3$ was an effective ligand of $Pd_2(dba)_3$ in a manner similar to the amination of aryl dibromides with secondary diamines. The M_w values were in the range of 9000–37 700.

Meta-substituted AB-type monomer such as *m*-bromoaniline also underwent Pd-catalyzed polycondensation [65], but the *para* counterpart has not been reported, possibly because aryl halides bearing the NH group in the *para* position are generally poor substrates for the aryl amination process. Buchwald et al. prepared a dimeric *para*-substituted AB monomer bearing the BOC protecting group and polymerized it with $Pd_2(dba)_3$ and 2-(di-*tert*-butylphosphino)biphenyl [73]. The polymerization carried out at room

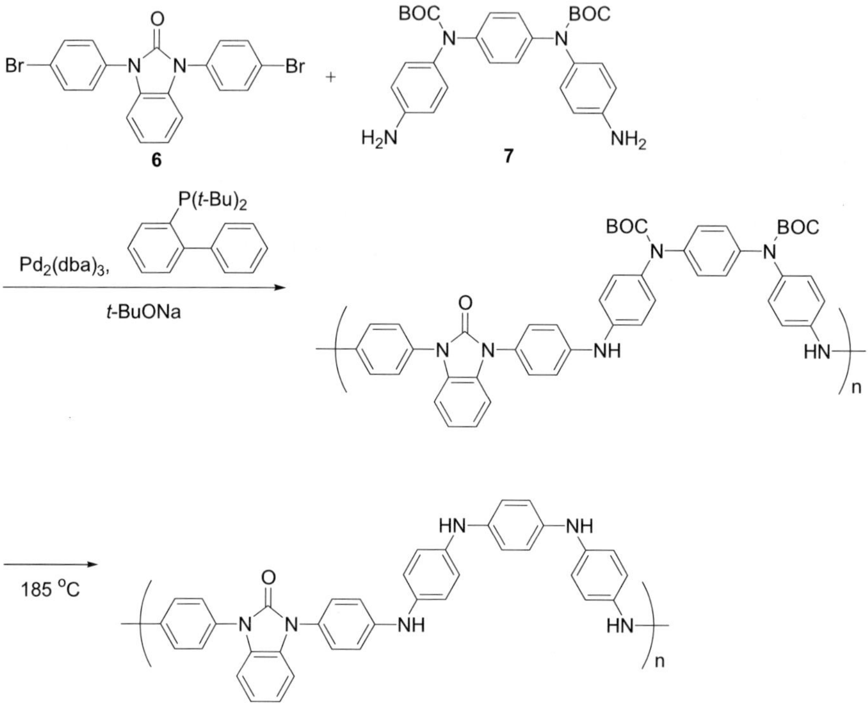

Scheme 15

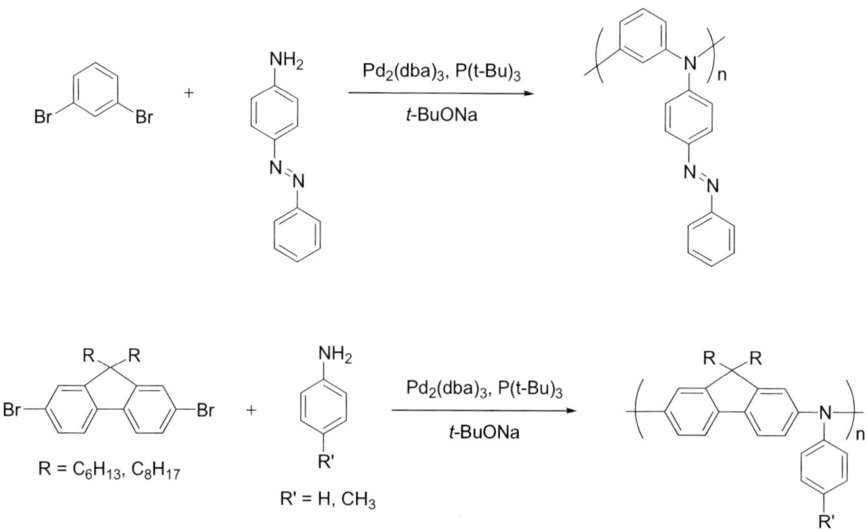

Scheme 16

temperature in THF using 1.0 mol % of $Pd_2(dba)_3$ and 3.0 mol % of the phosphine ligand gave the best result, with a M_w of 98 300 in 92% yield. This polymer is a BOC-protected polyaniline, which is soluble in common organic solvents and easily deprotected. Thermolysis of the polymer film at 185 °C under an inert atmosphere afforded polyaniline in the fully reduced LB form, whereas thermolysis in air affords the deep blue, partially oxidized EB form. Alternatively, the BOC-protected polymer film was directly transformed to the dark green, conductive ES form simply by immersion in acid solution under air (Scheme 17).

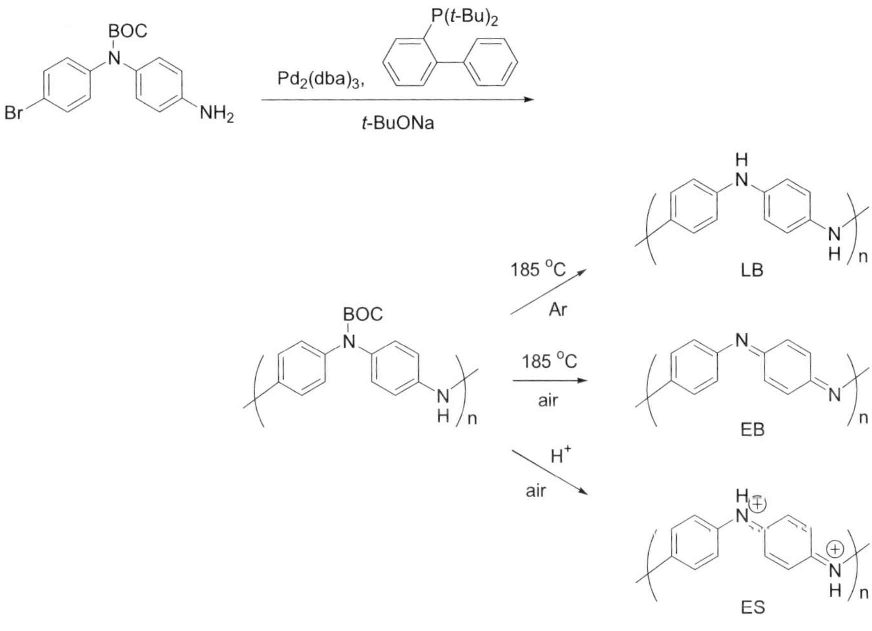

Scheme 17

Shibasaki and Ueda applied Pd-catalyzed etherification, developed by Buchwald et al. [74], to the synthesis of polyether [75]. Dibromobiphenyl underwent etherification with a bisphenol in the presence of 4 mol % of $Pd(OAc)_2$, 6 mol % of 2-(di-*tert*-butylphosphino)biphenyl, and 6 equivalents of K_3PO_4 at 90 °C (Scheme 18). The M_n value of the polymer was 18 000. Many aromatic polyethers have been synthesized by nucleophilic substitution of aromatic halides having electron-withdrawing groups, such as the sulfonyl group and keto group, with bisphenols. However, these electron-withdrawing groups in the backbone cause transetherification, resulting in limitation of molecular weight. The Pd-catalyzed polyetherification, in which aromatic dihalides without an electron-withdrawing group can be used, can afford new types of aromatic polyethers and does not contain this side reaction.

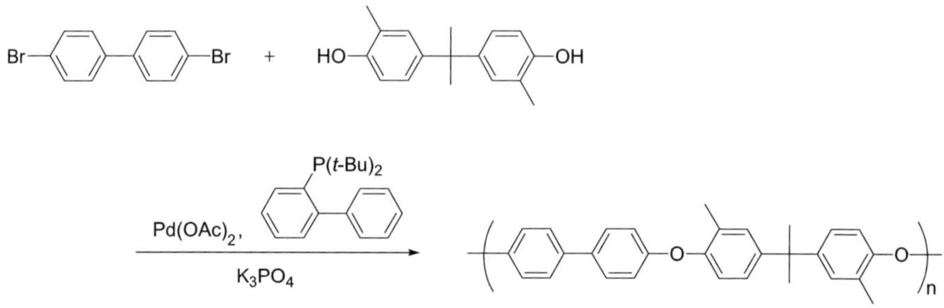

Scheme 18

The Pd-catalyzed synthesis of polycarbonate from bisphenol A and carbon monoxide has also been developed in the last decade (Scheme 19). Aromatic polycarbonates are generally prepared from toxic phosgene and bisphenol or its transesterification with carbonic acid derivatives such as diphenyl carbonate, which in turn are also prepared by phosgenation. In recent years, there has been increasing demand for a safer and more environmentally benign process for polycarbonate synthesis. Takeuchi and coworkers started investigation of direct polycondensation of bisphenol A with CO by using a $PdCl_2$–$Cu(OAc)_2$-based redox catalyst system [76]. The M_w of the obtained polymer was 3600. Analysis of the polymer end groups by MALDI-TOF mass spectrometry revealed that *o*-phenylene carbonate and salicylic acid structures were formed by oxidation of the terminal bisphenol A unit.

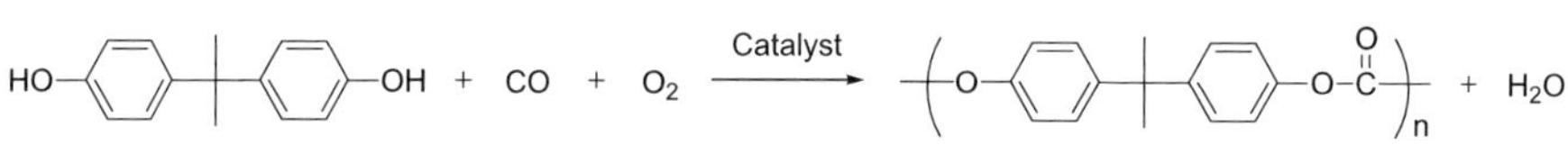

Scheme 19

To suppress the undesirable oxidation of the terminal group, various modifications were made in the catalyst system. On replacing $Cu(OAc)_2$ with $Ce(OAc)_3$, the *o*-phenylene carbonate terminal group was not formed, and the use of bis(triphenylphosphoranylidene) ammonium bromide instead of tetrabutylammonium bromide resulted in suppression of the salicylic acid end group. The M_w, however, was still 3800. Since the obtained polymer had the hydroxy groups at both chain ends, the polycondensation of this polymer, instead of bisphenol A, and CO was again carried out under the same conditions to obtain high molecular weight polycarbonate with M_w of 33 000 [77]. These results imply that the Pd catalyst was inactivated during polymerization. Finally, Takeuchi and coworkers found that the Pd/2,2′-bipyridyl complex with 6,6′-dimethyl groups (Scheme 20a) [78] and the Pd/carbene complex (Scheme 20b) [79, 80] gave polycarbonate with M_w of 12 900 and 24 000, respectively. The Pd/carbene complex anchored to a polystyrene resin (Scheme 20c) was also similarly effective; the total turnover number was 5100 [81].

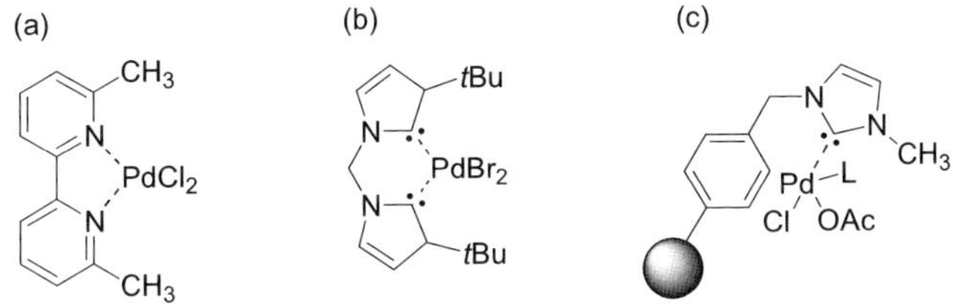

Scheme 20

Poly(silyl ether)s were classically synthesized by polycondensation of dichlorosilanes and diols, in which high molecular weight polymers were not obtained [82, 83]. In the last decade, however, efficient synthesis of poly(silyl ether)s by Pd- or Rh-catalyzed polycondensation with dehydrogenation has been reported. Li and Kawakami polymerized 1,4-bis(dimethylsilyl)benzene with labile hydrogen-containing compounds, such as water, ammonia, disilanol, aliphatic and aromatic diols, and dicarboxylic acids, in the presence of $Pd_2(dba)_3$ or $RhCl(PPh_3)_3$ at 25–50 °C (Scheme 21) [84–86]. The polymerization proceeded exothermically with vigorous gas evolution. The poly(silyl ether) obtained by the reaction with water possessed the M_n of 163 000 and formed a spherulite texture from melt (T_m = 123.5 °C). Mark et al. also conducted the dehydrogenative polycondensation of 1,4- or 1,2-bis(dimethylsilyl)benzene and 1,4- or 1,3-(dimethylhydroxysilyl)benzene with $RhCl(PPh_3)_3$ catalyst [87]. The poly(silphenylenesiloxane) with the highest M_n of 17 200 was obtained from both 1,4-substituted monomers (Scheme 22).

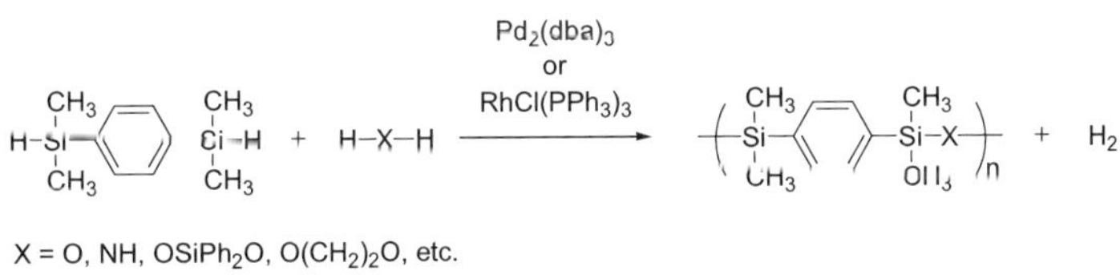

Scheme 21

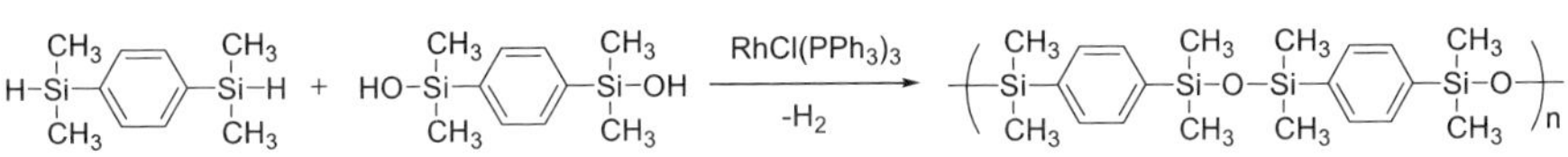

Scheme 22

These polysiloxanes were also found to be synthesized by the condensation reaction between organosilanes and organoalkoxysilanes with release of alkane as an inert by-product. For this polycondensation, $B(C_6F_5)_3$ was used as an effective catalyst [88]. This process involves cleavage of C–O and Si–H bonds and formation of Si–O and C–H bonds, and hence the reaction is extremely exothermic due to a very large enthalpy of reaction, $\Delta H \approx -250\,\mathrm{kJ\,mol^{-1}}$. The experiment should be carefully carried out. It is preferable to add the dialkoxysilane to a mixture of the hydrosilane and catalyst. The dehydrogenative polycondensations, mentioned above, use 0.2–0.5 mol % of Pd and Rh catalysts, whereas this polycondensation effectively proceeds with only 0.004 mol % of the catalyst, and high molecular weight materials (M_w = 10 000–50 000) were obtained (Scheme 23).

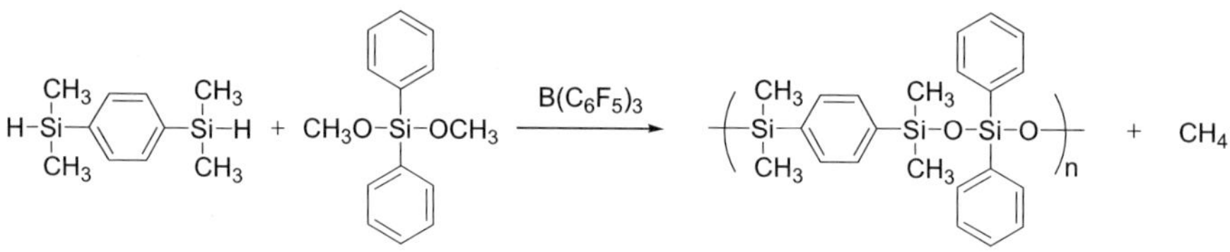

Scheme 23

3 Control of Chemoselectivity

Chemoselective condensation polymerization is applicable to the synthesis of sequential condensation polymers from unsymmetrical monomers and of polymers with pendant functional groups without using protecting groups. The former study had been conducted since the 1960s, and Ueda summarized it by including his work in a review about a decade ago [89]. Ueda et al. also initiated the latter study in 1988 and synthesized polyamides containing hydroxy, amino, and carboxyl groups on the aromatic rings of the polymer backbone by the chemoselective polyamidation of dicarboxylic acids with diamines bearing these functional groups in the presence of condensing agent, DBOP [90, 91]. Wang and Nakamura synthesized polyesters containing carboxyl group, quaternary ammonium, sulfonic acid, and chlorohydrin moieties by the chemoselective polycondensation of isophthaloyl dichloride and these functional diols [92–97].

More elaborate chemoselective polycondensations have been reported since 2000. Ueda and coworkers tried direct polycondensation of 5-hydroxy or 5-aminoisophthalic acid and an aromatic diamine bearing a phenol moiety in the presence of DBOP (Scheme 24) [98]. The model reactions were first carried out to find the conditions for chemoselective polycondensation. The reaction of 5-hydroxyisophthalic acid with *p*-toluidine in the presence of DBOP and triethylamine afforded the corresponding amide in a way similar to ordinary amidations with DBOP. On the other hand, the reaction of 5-aminoisophthalic acid with *p*-toluidine was dependent on the base used. When triethylamine, DBN, and DBU were used, a mixture of the desired

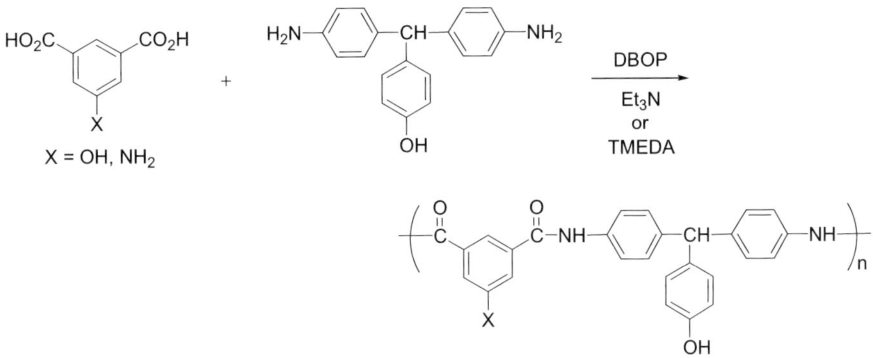

Scheme 24

amide and a hyperbranched polymer from 5-aminoisophthalic acid were obtained. When TMEDA, a slightly weaker base, was used, selective amidation with *p*-toluidine was attained without reaction of the amino group of 5-aminoisophthalic acid. On the basis of these results, the polycondensations of 5-hydroxyisophthalic acid with the diamine and of 5-aminoisophthalic acid with the diamine were carried out with triethylamine and TMEDA in the presence of DBOP, respectively. The polyamide obtained in the former polycondensation showed an inherent viscosity of $0.73\,dL\,g^{-1}$, and the latter polyamide $0.48\,dL\,g^{-1}$. However, the ^1H NMR spectrum of the latter polyamide showed that about 3% of the pendant amino group was converted into amide linkage.

Furthermore, the chemoselective polyesterification of isophthalic acid with a bisphenol bearing amino group was investigated by direct polycondensation with DBOP and base (Scheme 25) [99]. Ueda and coworkers expected that phenolate was more nucleophilic than the aromatic amino group,

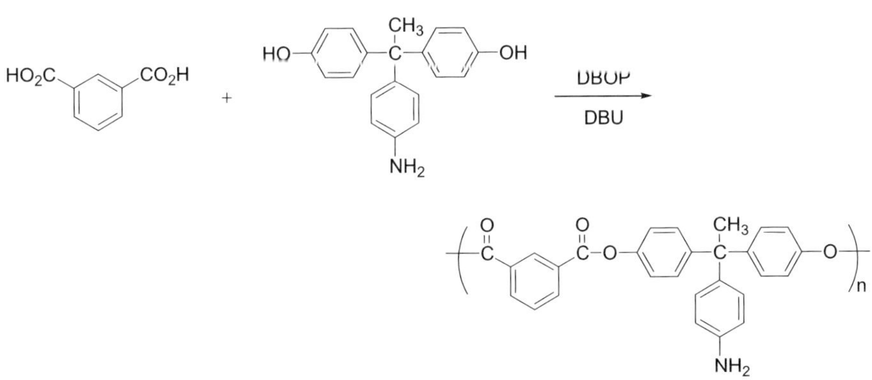

Scheme 25

although phenol was less nucleophilic than the aromatic amino group. Accordingly, this polymerization was also dependent on the base used in the reaction. When triethylamine was used, amidation predominantly occurred. When DBN, a stronger base, was used, esterification took place but phenoxide also attacked DBOP to give phosphate. To suppress this side reaction, isophthalic acid was first reacted with DBOP in the presence of DBN to generate a carboxylic–phosphoric anhydride, then a solution of the bisphenol and DBN was added. The obtained polyester bearing the pendant amino group possessed a M_w of 22 000 and T_g of 192 °C.

Takasu et al. found that polycondensation catalyzed by $Sc(OTf)_3$ under mild conditions, mentioned in Sect. 2.2 and Scheme 5, can distinguish primary and secondary hydroxy groups of monomers, and it was applied to the chemoselective synthesis of polyesters bearing the secondary hydroxy group [100]. Malic acid and tartaric acid, having pendant secondary hydroxy groups, were reacted with 1,9-nonanediol in the presence of 0.5 mol% of $Sc(OTf)_3$ under reduced pressure (0.3–3 mmHg) at 80 °C to give polyesters with M_n of 13 100 and 8300, respectively, without reaction of the pendant

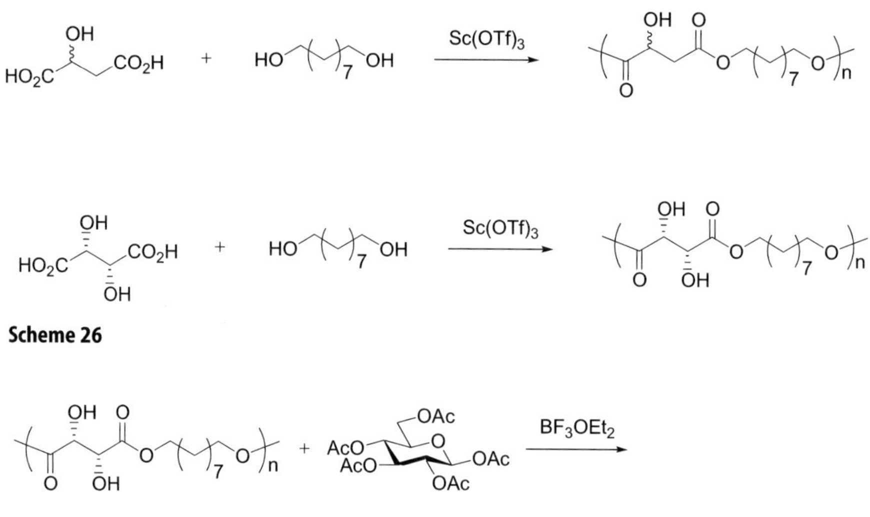

Scheme 26

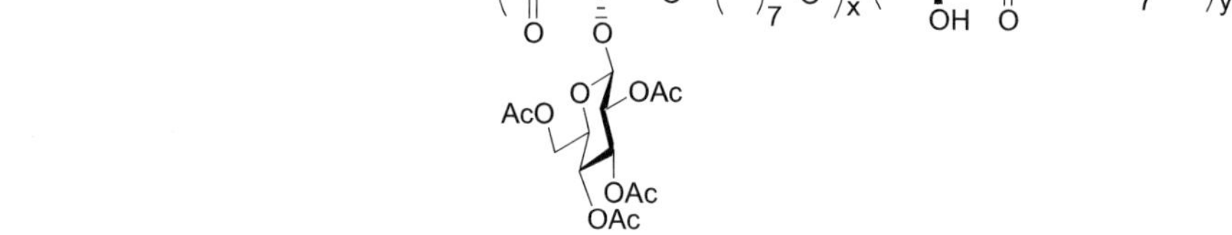

Scheme 27

secondary hydroxy groups in the dicarboxylic acids (Scheme 26). The polycondensation of diols containing secondary hydroxy groups, including glycerol and sorbitol, with dicarboxylic acids also proceeded chemoselectively to afford the corresponding polyesters. The obtained polyester smoothly underwent glycosidation by the reaction of the secondary hydroxy group on the polyester with a glycosyl donor by virtue of BF_3OEt_2 as a catalyst to afford a sugar-containing polyester (50% glycosidation) (Scheme 27). Lipase-catalyzed copolycondensation of malic acid, adipic acid, and 1,8-octanediol also gave polyester bearing hydroxy groups, the M_w of which was dependent on the reaction solvent and increased up to 17 800 by polymerization in hexane [101, 102].

4 Control of Stereochemistry

4.1 Regioselective Polymerization

4.1.1 General Polycondensation Between Nucleophile and Electrophile

Regioselective polycondensation of polyols, such as sugars, reduced sugar polyols, and glycerol, has been generally achieved by enzyme catalysts except for a recent work of $Sc(OTf)_3$-catalyzed polyesterification of polyols mentioned in the previous section [100]. Kobayashi et al. reported the first synthetic cellulose by polycondensation of β-D-cellobiosyl fluoride, a monomer having seven hydroxy groups and one fluorine, with cellulase, a hydrolysis enzyme of cellulose, in the early 1990s [103]. After that, xylan [104], amylose [105], a cellulose–xylan hybrid [106], and an unnatural polysaccharide, an alternating 6-*O*-methylated cellulose [107], were synthesized by polycondensations of the corresponding sugar fluorides catalyzed by hydrolase enzymes. On the other hand, glycosaminoglycans, such as hyaluronan and chondroitin, were synthesized by enzymatic ring-opening polyaddition, not polycondensation, of sugar oxazoline-type monomers [108].

Regarding polyesters, Russell and coworkers reported the lipase (Novozyme-435)-catalyzed polycondensation of various triols and divinyl adipate for the synthesis of polyesters having pendant hydroxy groups (Scheme 28) [109]. Before this report, the divinyl adipate was reported to be available for enzymatic polyesterification of diols by Kobayashi and coworkers [110, 111], in which the equilibrium of esterification is shifted in the forward direction by tautomerization of the eliminated vinyl alcohol to acetaldehyde. The M_w values of the resulting polyesters varied according to the triol used and ranged from 3000 to 14 000. Analysis by MALDI-TOF mass

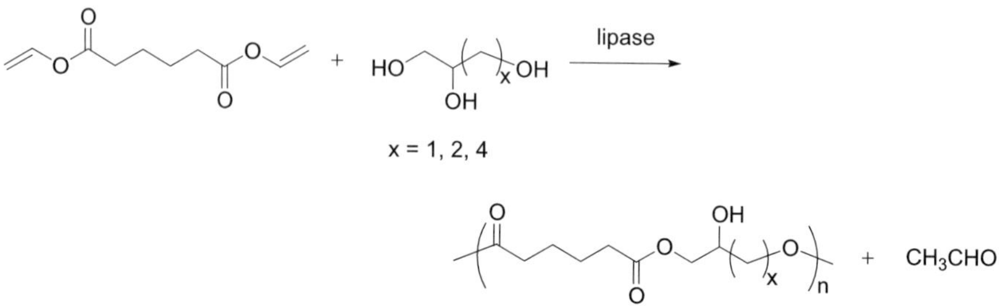

Scheme 28

spectrometry showed the presence of a linear polyester with hydroxy substituents, and there was no evidence for network formation. The pendant groups were 90–95% secondary and 5–10% primary hydroxy groups. Although the activation of carboxylic acids with electron-withdrawing groups had been thought to be necessary for enzyme-catalyzed polyesterifications of polyols [112–120], Gross and coworkers recently reported the direct lipase-catalyzed condensation polymerization between diacids and reduced sugar polyols [121]. The polycondensation of adipic acid and sorbitol was performed in bulk for 48 h at 90 °C using Novozyme-435. The M_w of the obtained polyester was 17 030, and sorbitol was esterified with high regioselectivity (85 ± 6%) at the primary hydroxy groups. (Scheme 29). When adipic acid, 1,8-octanediol, and sorbitol were copolymerized in the molar ratio 50:35:15, the methanol-insoluble product had a M_w of 117 000.

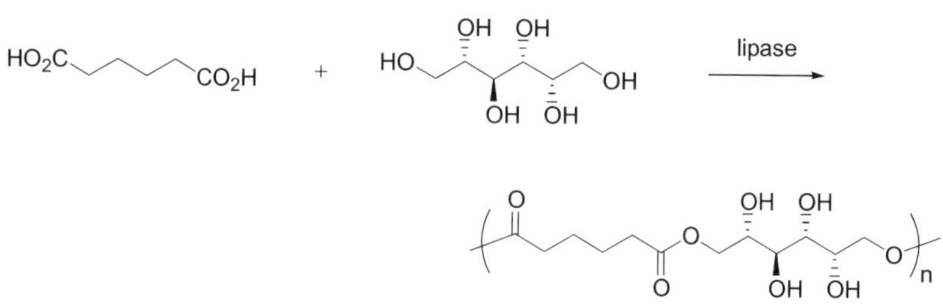

Scheme 29

Regioselective polycondensations with transition-metal catalysts were also reported. Nomura et al. developed palladium-catalyzed allylation polycondensation, in which nucleophile predominantly reacted with π-allyl palladium at the terminal allylic carbon to give *E*-linear products [122, 123]. On the other hand, polymerization with an iridium catalyst selectively proceeded at the internal allylic carbon to yield branched polymers with pendant vinyl groups (Scheme 30). These polycondensations demonstrate that polymers having different structures can be synthesized from the same monomers by changing the catalyst [124].

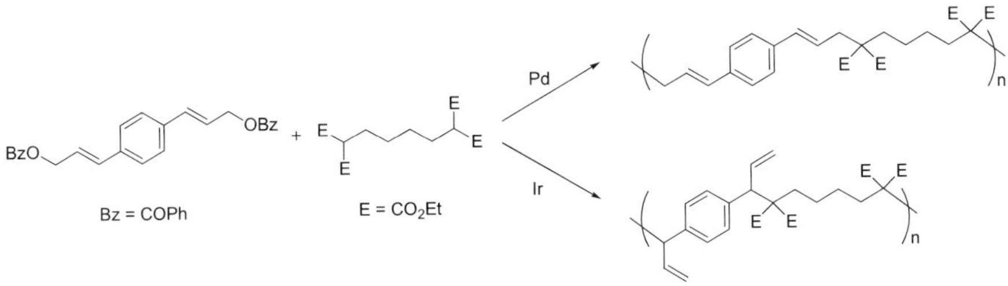

Scheme 30

4.2
Oxidative Polycondensation

In oxidative polymerization of phenol and naphthol derivatives, the structure of the resulting polymers is often very complicated since they are multifunctional monomers. Only some 2,6-disubstituted phenols are oxidatively polymerized to give poly(1,4-oxyphenylene)s exclusively [125, 126]. Even enzymatic polymerization of phenol derivatives with peroxidase afforded polyphenols consisting of a mixture of phenylene and oxyphenylene units, which are formed by the C–C and C–O coupling of phenols, respectively [127–129]. In the enzymatic polymerization of 4-substituted phenols, however, the regioselectivity was controlled by changing the hydrophobicity of the monomer and solvent in the range of the phenylene content from 40 to 70% [130, 131]. Furthermore, Kobayashi and coworkers found that addition of PEG to the reaction mixture for the enzymatic polymerization of phenol resulted in the polymer mainly consisting of the phenylene unit (95%) [132]. The role of PEG was thought to be a template for the oxidative polymerization of phenol, because the hydrogen-bonding interaction of PEG with the hydroxy group of phenol and polyphenol was observed by UV absorption and FT-IR spectroscopy (Scheme 31).

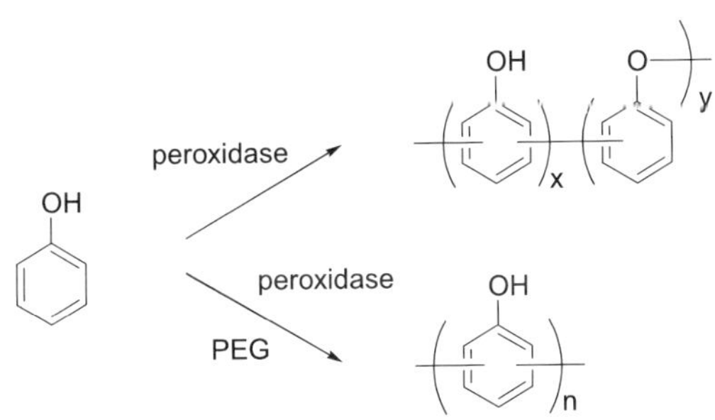

Scheme 31

On the other hand, polyphenol consisting of oxyphenylene units, PPO, was successfully synthesized by the highly regioselective oxidative polycondensation of 4-phenoxyphenol with tyrosinase model complexes [133, 134]. Higashimura, Moro-oka, Kobayashi, and coworkers selected two copper(II) chloride complexes, Cu(tpzb)Cl [135] and Cu(tacn)Cl_2 [136], as catalysts for the oxidative polycondensation. The polymerization mechanism was postulated as follows (Scheme 32). First, Cu(tpzb)Cl and Cu(tacn)Cl_2 react with 4-phenoxyphenol or its oligomers to give phenoxo copper(II) complex **8**, equivalent to phenoxy radical–copper(I) complex **9**. Intermediates **8** and/or **9** are not free radicals but "controlled" radicals, and hence regioselective coupling takes place to produce copper(I) complexes **10** of Cu(tpzb) and Cu(tacn)Cl. The complexes **10** afford μ-η^2:η^2-peroxo dicopper(II) complex **11** [137], a copper–dioxygen model complex for tyrosinase, under dioxygen and react with phenols to regenerate intermediates **8** or **9** and hydrogen peroxide. The coupling between the phenoxy radicals was thought to proceed via a quinone-ketal intermediate, which would be subjected to rearrangement to afford the linear oligomer (Scheme 33). The polymers obtained showed the same IR spectra as crystalline PPO, and their melting points were 171–194 °C, which were lower than that of linear, high molecular weight PPO

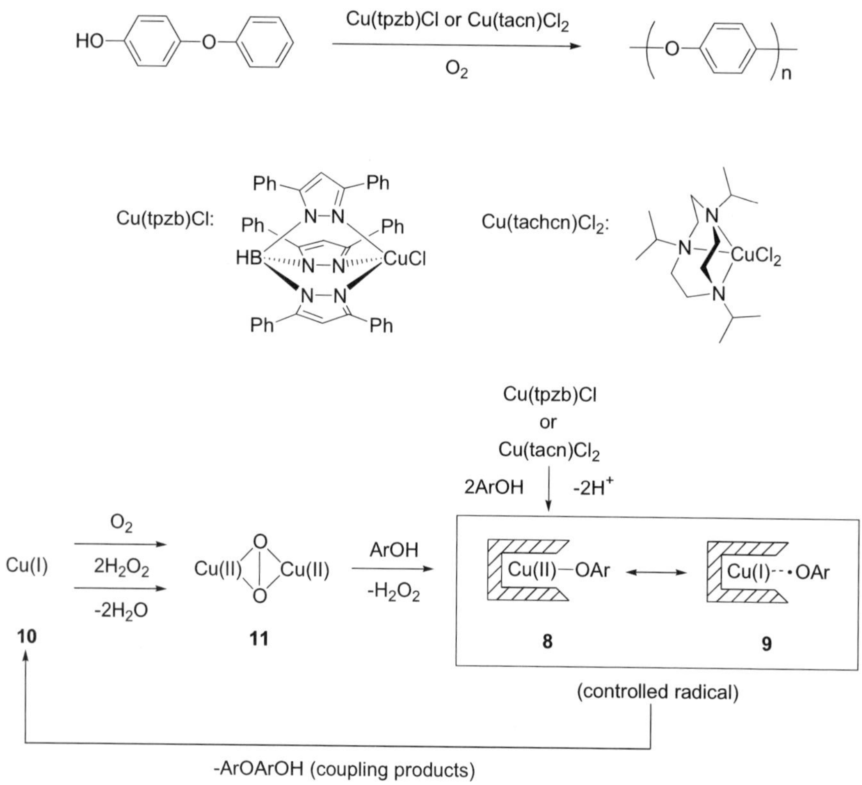

Scheme 32

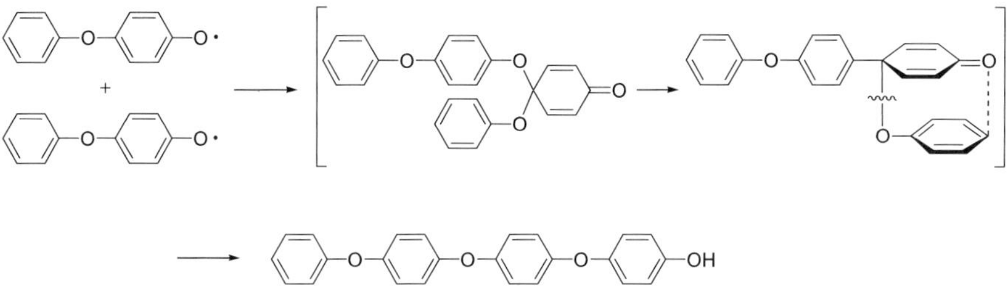

Scheme 33

(298 °C) [138]. This may be because the polymers contained a small amount of *ortho* C–O linkage (<9%) and/or had a lower molecular weight.

The oxidative polycondensation of 2,5-dialkylphenols was also conducted by this polymerization system [139]. In the polymerization of 2,5-dimethylphenol with 5 mol % of Cu(tacn)Cl_2, the M_w was 19 300 and the polymer was composed exclusively of a 2,5-dimethyl-1,4-phenylene oxide unit (Scheme 34). The DSC traces of the polymer showed heat-reversible crystallinity with a high melting point at ca. 300 °C. The isomeric polymer 2,6-dimethyl-PPO showed a melting point at 237 °C [140]. However, once the crystal part had been totally melted, recrystallization never occurred during slow cooling or after annealing in the temperature range between T_g (209 °C) and melting point [140]. In the oxidative polymerization of the diethyl and dipropyl counterparts, the M_w values of the obtained polymers were 23 100 and 32 200, respectively. Interestingly, the latter 2,5-dipropyl-PPO possessed heat-reversible crystallinity, whereas the former 2,5-diethyl-PPO did not. Shibasaki and Ueda et al. also conducted the regioselective oxidative polycondensation of 2,5-dimethylphenol by a copper-amine catalyst immobilized on mesoporous silica material (Scheme 34) [141]. They proposed that regioselective polymerization was induced by the interior of the mesoporous channel (6.0 nm) because the polymerization with conventional CuCl-amine catalyst showed poor regioselectivity. In the polymerization of *o*-cresol with the same catalyst, a regioselectivity of 80% was attained in the presence of 2-phenylpyridine [142].

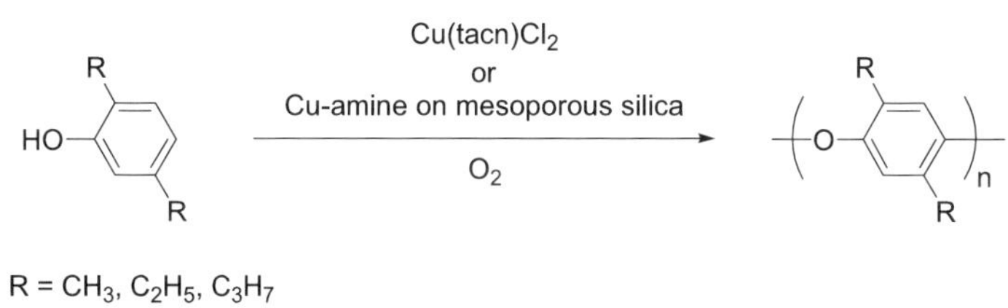

Scheme 34

Not only phenol derivatives but also 1,4-dialkoxybenzenes undergo oxidative polycondensation to give poly(*p*-phenylene). However, the oxidative polycondensation using excess amounts of $FeCl_3$ as an oxidant and Lewis acid showed no regioselectivity; almost 1:1 of 1,3- and 1,4-linkages [143]. Ueda and coworkers found that the use of $VO(acac)_2$ catalyst [144] under an oxygen atmosphere for the oxidative polycondensation of 1,4-dibutoxybenzene resulted in the almost completely 1,4-regiocontrolled polyphenylene, poly(2,5-dibutoxy-1,4-phenylene) (Scheme 35) [145]. The M_n was 10 000–12 000. The maximum absorption wavelength of the regioregular polymer (λ_{max} = 331–333 nm) was slightly higher than that of the regiorandom polymer obtained with $FeCl_3$ (λ_{max} = 328 nm).

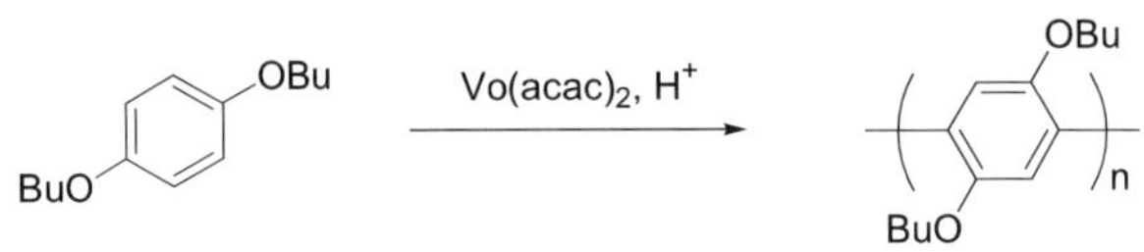

Scheme 35

The oxidative coupling of phenol derivatives involves mainly C–O coupling as mentioned above, whereas that of naphthol derivatives involves C–C coupling. For example, Koga and coworkers reported that the oxidative coupling of 2-naphthol with CuCl(OH)TMEDA catalyst selectively afforded 1,1′-bi(2-naphthol) [146]. Ueda and coworkers applied this reaction system to the regioselective polycondensation of a naphthol derivative, which was prepared from 6-bromo-2-methoxynaphthalene and bisphenol A in two steps [147]. The oxidative polymerization of the monomer was carried out in the presence of 5 mol% of CuCl(OH)TMEDA at room temperature to yield a polymer with the M_w of 33 000 (Scheme 36). The polymer showed a good thermal stability; a 10% weight loss occurred at 450 °C. A simple monomer, 2,6-dihydroxynaphthalene, also polymerized with this catalyst to give a polymer with the M_w of 93 600 [148]. The polymer was confirmed to consist completely of the 1,5-linkage from ^1H and ^{13}C NMR analysis (Scheme 37). This catalyst, however, was not effective for the oxidative coup-

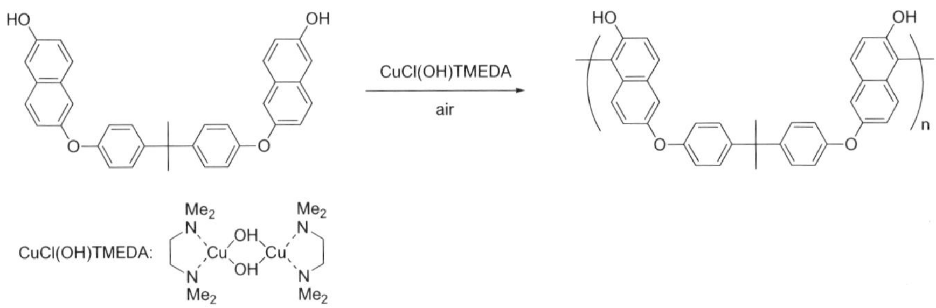

Scheme 36

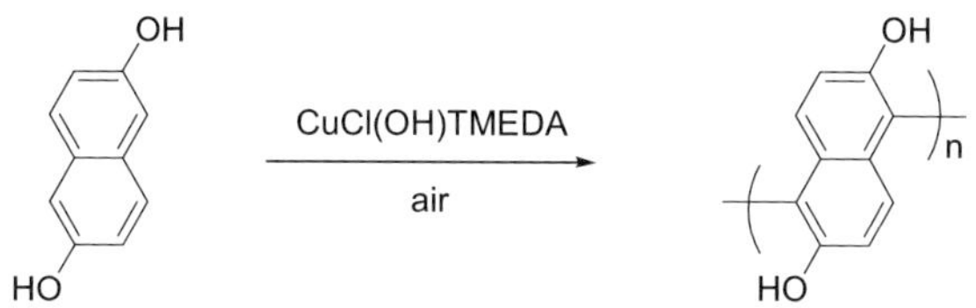

Scheme 37

ling of 1- and 2-alkoxynaphthalenes, the oxidation potentials of which are higher than that of 2-naphthol [147]. Accordingly, the oxidative polymerization of a 1-naphthyl ether monomer was conducted with 2.5–5.0 equivalents of $FeCl_3$ [149]. The coupling reaction selectively took place at the 4-position of the 1-aryloxynaphthyl moiety (Scheme 38). This polymer possessed a M_w of 44 600 and was extremely thermally stable; the 5% weight loss temperature and T_g were 520 and 301 °C, respectively. Furthermore, the molecular density was so low that the polymer showed a low dielectric constant of 2.50. This polymer will be a good insulating material for future generations of microchips.

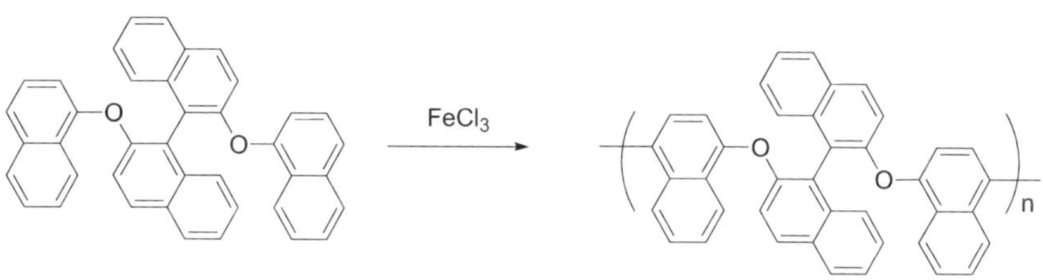

Scheme 38

If aromatic monomers polymerize with high regioselectivity, the polymerization of monomers bearing the substituent at the dissymmetrical position gives rise to another problem of stereochemistry: regioregularity. For example, in oxidative polymerization of 3-alkylthiophenes, regioselectively proceeds at the 2- and 5-positons of the thiophene ring, but the polymer includes four types of diad sequences: head-to-tail (HT), head-to-head (HH), tail-to-tail (TT), and tail-to-head (TH) (Scheme 39). Anderson et al. reported the synthesis of HT poly[3-(4-alkylphenyl)thiophene] with an excellent regioregularity through the oxidative polycondensation of the corresponding thiophene by slow addition of $FeCl_3$ [150]. Ueda and coworkers also attained high regioregularity by the use of a catalytic amount of $VO(acac)_2$ [151]. Regarding poly(3-alkylthiophene), Wegner and coworkers reported that 80% of HT regioregularity was obtained by oxidative polymerization with $FeCl_3$ in a dilute solution [152], and Ueda and coworkers reported 88% of HT regioregularity under the optimum conditions [153]. Besides oxidative polycondensation, McCullough et al. [154] and Rieke et al. [155] achieved more

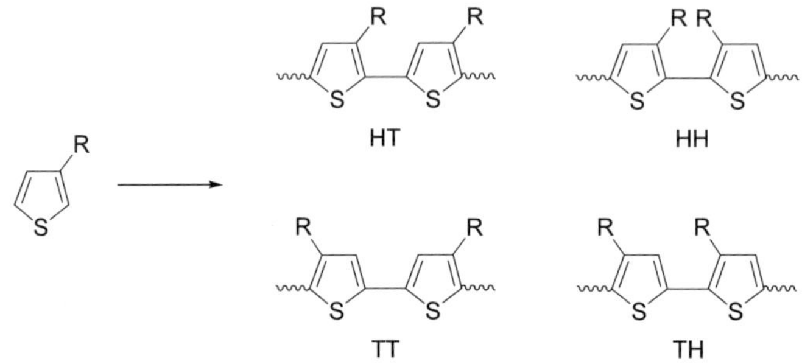

Scheme 39

than 95% of HT regioregularity of poly(3-alkylthiophene)s by the polymerization of 3-alkyl-5-bromo-2-metallated thiophenes with a Ni catalyst.

4.3 Geometrically Selective Polymerization

PPVs have applications in optoelectronic devices, such as light-emitting diodes and photovoltaic cells, and the geometries of the vinylene linkages are known to profoundly affect the optical properties [156–160]. However, efficient ways of regulating these geometries had remained almost unexplored. Katayama and Ozawa et al. recently synthesized all-*trans*-[161] and all-*cis*-PPVs [162, 163] by using stereocontrolled hydrosilylation of *p*-diethynylbenzene with ruthenium catalysts (Scheme 40). Thus, $RuHCl(CO)(PPh_3)_3$ formed (*E*,*E*)-isomer, whereas $[RuCl_2(CO)(DPCB\text{-}OMe)]_2$ afforded (*Z*,*Z*)-isomer [164] with over 99% selectivities in both cases. The (*E*,*E*)-isomer underwent Hiyama coupling polycondensation with 2,5-dioctyloxy-1,4-diiodobenzene in the presence of a palladium catalyst and TBAF hydrate, giving all-*trans*-PPVs with more than 99% geometrical purity [161]. In the Hiyama coupling polycondensation of the (*Z*,*Z*)-isomer with the diiodobenzene, however, the stereoregularity for *cis*-PPV was limited to 66%. The reason for this drop in stereoregularity is that *trans* isomer is formed by TBAF-induced desilylation of the alkenylsilane followed by Mizoroki–Heck arylation of the resulting alkene. Probably, the bulky octyloxy groups on the diiodobenzene retard the transmetalation of the (*Z*,*Z*)-isomer with an arylpalladium complex formed by oxidative addition of the diiodobenzene to a palladium(0) species, resulting in the TBAF-induced desilylation of the (*Z*,*Z*)-isomer [163]. Overcoming this side reaction, they converted the (*Z*,*Z*)-isomer into the corresponding (*Z*,*Z*)-bromide with *N*-bromosuccinimide, then performed Suzuki–Miyaura coupling polycondensation of this bromide and diboronic acid in the presence of a palladium catalyst. The obtained PPV was confirmed to be all-*cis* by NMR spectroscopy [162]. The all-*cis*-PPVs underwent one-way photoisomerization to

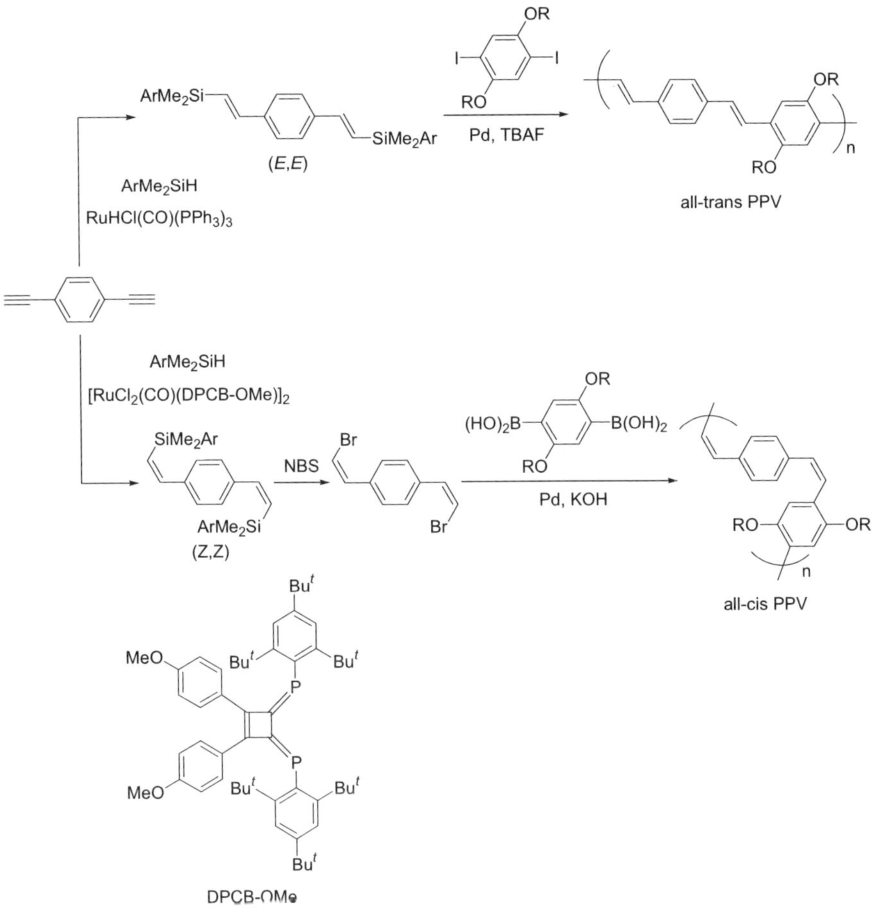

Scheme 40

the corresponding *trans*-PPVs, both in solution and in the solid. This phenomenon was applied to direct microscale patterning of PPVs onto a quartz substrate [162, 165].

4.4
Enantioselective Polymerization

Asymmetric oxidative coupling polymerization of hydroxynaphthalene derivatives was investigated by Habaue and Okamoto et al. First, they studied the oxidative coupling polymerization of optically active 3,3′-hydroxy-2,2′-dimethoxy-1,1′-binaphthalene with copper catalysts bearing chiral ligands under an oxygen atmosphere (Scheme 41) [166]. The obtained polymers had molecular weights of 3100–5200. When the polymerization of (*R*)-monomer

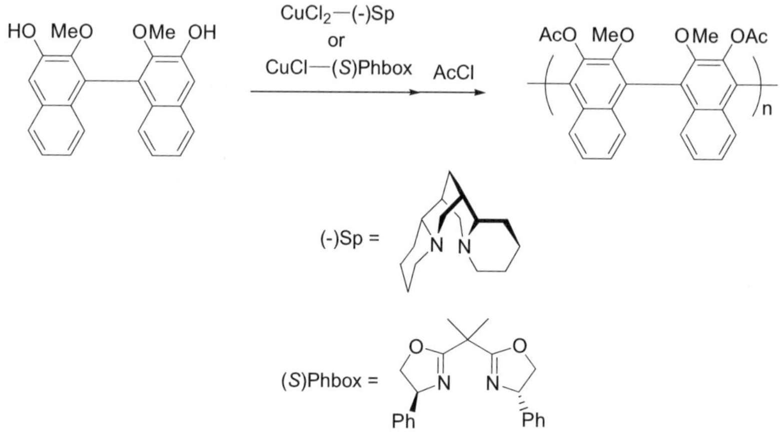

Scheme 41

was carried out with $CuCl_2$-(–)-sparteine complex, the selectivity was $R:S=84:16$ (68% ee). The polymerization of (*S*)-monomer with the same catalyst, however, showed almost no selectivity ($R:S=52:48$). Another catalyst, CuCl-(*S*)-Phbox complex, also showed similar selectivity (52–60% ee) [167]. The polymerization of the MOM counterparts instead of the methoxy group with CuCl-(*S*)-Phbox complex showed 52% ee [168].

They next studied the asymmetric oxidative polymerization of achiral 2,3-dihydroxynaphthalene (Scheme 42). The polymerization of this monomer with $CuCl_2$-(–)-sparteine complex resulted in a low yield and gave a low molecular weight oligomer, whereas the polymerization with CuCl-(*S*)-Phbox quantitatively gave a polymer with M_n of 10 600–15 300. The enantioselectivity attained in this polymerization, however, was estimated to be low, with 43% ee from the model reaction [169]. When vanadyl sulfate ($VOSO_4$)-Phbox complex was used instead of the copper catalyst system, the enantioselectivity was improved up to 80% ee [170]. Asymmetric cross-coupling polymerization of two kinds of naphthol derivatives was also reported [171, 172].

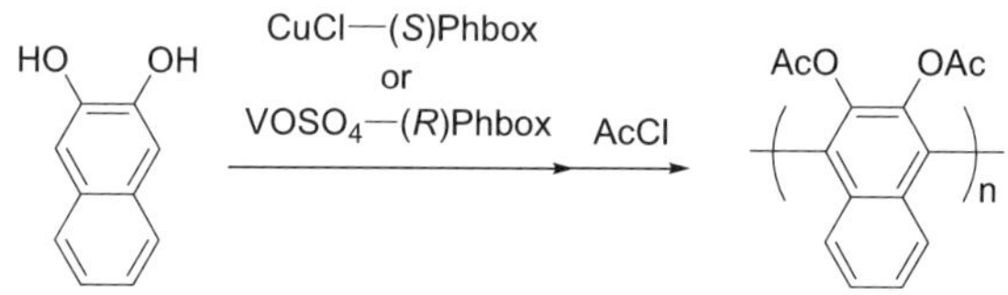

Scheme 42

Kawakami and coworkers synthesized stereoregular and optically active polysiloxanes by polycondensation with deamination or dehydrogenation. Optically active disiloxane disilanol [173] and achiral bis(dimethylamino)silane were polymerized in toluene at 100 °C to afford optically active polysiloxane with 100% retention of configuration (Scheme 43) [174]. The M_n of the polymers was 18400–170000. When tris(dimethylamino)silane was used instead of bis(dimethylamino)silane, optically active poly(silsesquioxane) gel was obtained [175]. If a chiral aminosilane was reacted with a chiral silanol, racemization of the silicon atom of the aminosilane seemed to occur [176]. Similarly, in the dehydrogenative polycondensation of the above disilanol and optically active disiloxane with an Rh catalyst at 60 °C, racemization of the silicon atom of the disiloxane took place (Scheme 44) [176]. When an achiral silane was used, the chirality of the silicon atom of the disilanol was confirmed to be retained [177].

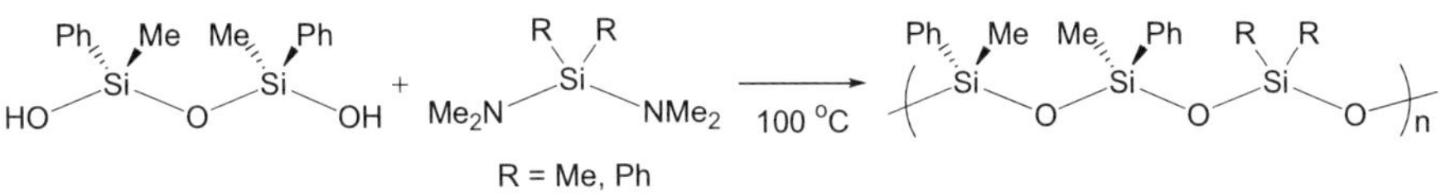

Scheme 43

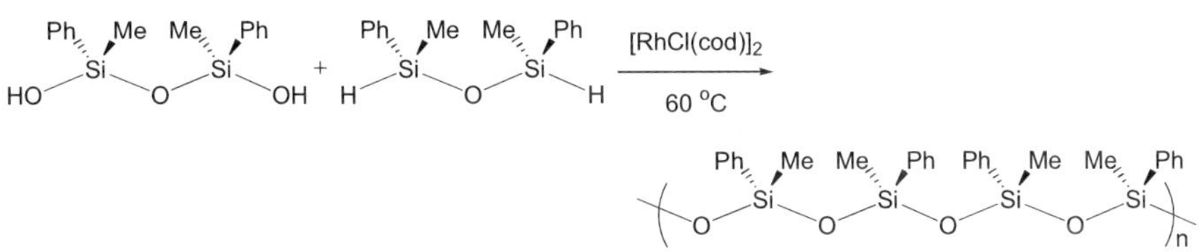

Scheme 44

Optically active poly(silyl ether)s were also synthesized by asymmetric induction on the silicon atom from chiral diols or a chiral Rh catalyst [178]. In the dehydrogenative polycondensations of bis(silane)s and several chiral diols, the highest ee value of the silicon atom was 13.7%. In the polymerization of a bis(silane) and an achiral diol, however, the use of only 5 mol% of RhCl[(*R*) BINAP] resulted in induction of 39.8% ee of the silicon atom (Scheme 45).

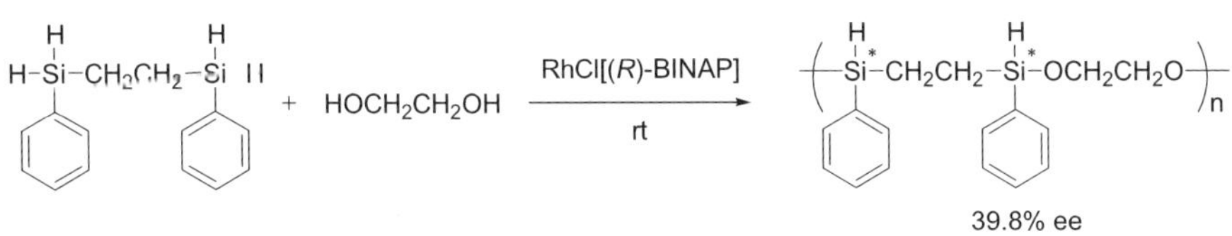

Scheme 45

5
Control of Chain Topology

5.1
Synthesis of Cyclic Polymers

It is thought to be difficult to synthesize cyclic polymers by general polymerization methods, such as addition polymerization and ring-opening polymerization. Regarding polycondensation, it also seemed difficult, as indicated by the fact that Carothers and Flory did not take into account cyclization reactions in their theory of step-growth polymerization. However, it was identification of cyclic polymers that was difficult. The development of MALDI-TOF mass spectrometry during the past decade has provided a new powerful tool for qualitative analyses of cyclic oligomers and polymers.

Kricheldorf et al. first studied the reaction of tin-containing macrocyclic polylactone or copoly(ether–ester) with dicarboxylic dichloride to obtain polycondensates under elimination of dibutyltin chloride [179, 180]. They found limitation of the molecular weight of the obtained polymers and speculated that this result might come from competition between the propagation and cyclization steps [180]. In the polycondensation of 2,2-dibutyl-2-stanna-1,3-dioxepane and dicarboxylic dichloride, it was revealed that macrocyclic polyesters were the main products by means of MALDI-TOF mass spectrometry (Scheme 46) [181]. When acyclic 1,4-bis(tributylstannoxy)butane was reacted with dicarboxylic dichlorides, macrocyclic polyesters were preferentially formed. Therefore, cyclic tin monomer did not play a decisive role in the formation of the macrocyclic polyesters. At this time they did not conclude that special interactions between the alkoxytin group and the acid chloride moiety at both ends of the polymer favored cyclization, or that the cyclization was a normal result of all polycondensations at high conversion. After this report, however, they demonstrated for numerous polycondensations that cyclization takes place at any concentration and at any stage of the polycondensation process. Therefore, cyclization limits the degree of polymerization (DP) as expressed by Eq. 1, where V_p and V_c are the rates of propagation and cyclization, and X is a constant >1.0 [182].

$$\mathrm{DP} = \frac{1}{1 - p\left(1 - \frac{1}{X^{\alpha}}\right)}\,; \qquad \alpha = V_p / V_c\,. \tag{1}$$

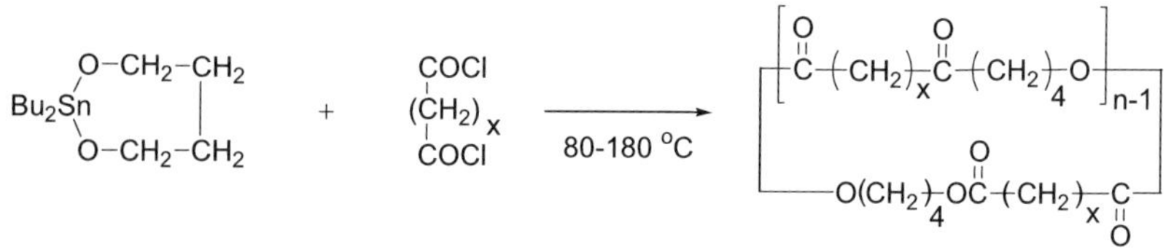

Scheme 46

Furthermore, Kricheldorf made the following statement: "All step-growth polymerizations possess the fundamental tendency to yield cycles as stable end products with linear chains being the reactive intermediates or end products of side reaction" [183].

Several examples of producing cyclic polymers are described below. Interfacial hydrolytic polycondensation of bisphenol A bischloroformate afforded polycarbonate with the M_n of 65 000 under optimized conditions (Scheme 47). The MALDI-TOF mass spectrum of this polymer displayed peaks of cyclic oligo- and polycarbonates up to 18 000 Da and no peaks for linear chains. Furthermore, the optimum mass spectrum of the polymer sample fractionalized by SEC showed the peaks of cycles up to 55 000 Da corresponding to a DP of around 210 [184]. This result indicates that cyclization competes with propagation at any chain length. It should be noted that the molecular weight resulting from this interfacial polycondensation was by factors of 12–20 higher than those obtained under homogeneous conditions. This result is also interpreted in terms of different efficiency of cyclization in the interfacial polycondensation. Under homogeneous conditions the reaction mixture of an $A_2 + B_2$ polycondensation contains three species with respect to the end group combination: "A–A chains", "B–B chains", and "A–B chains". Under ideal conditions, their molar ratio is 1:1:2. Since the A–A and B–B chains can contribute to the chain growth but not to cyclization, 50 mol% of the A–B chains are able to cyclize. In the case of the hydrolytic interfacial polycondensation of bisphenol A bischloroformate, the phenoxide ions, which are gradually generated by the hydrolysis of the chloroformate groups, are surrounded by a huge excess of chloroformate groups, so that the chains having two chloroformate groups largely dominate over the chains with the phenoxide and chloroformate groups. Therefore, the fraction of chains that can cyclize is far lower than in the case of a one-phasic polycondensation [185].

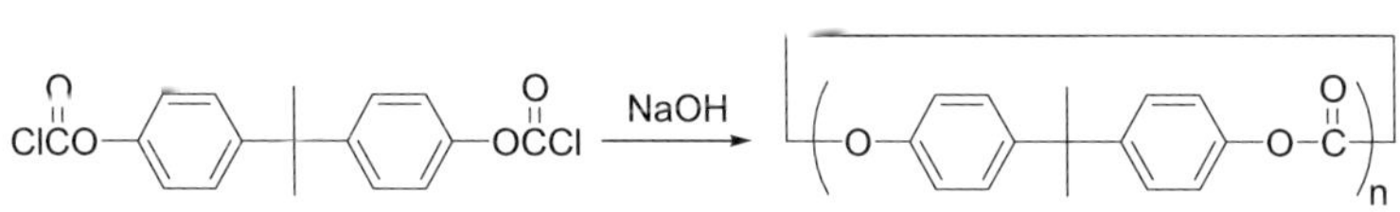

Scheme 47

Poly(ether sulfone)s are like polycarbonates, a group of engineering plastics, and are synthesized generally by the polycondensation of a bisphenol with 4,4′-dichlorodiphenylsulfone in the presence of K_2CO_3 in DMSO (Scheme 48). Commercial poly(ether sulfone)s have M_n in the range of 15 000 to 25 000, and the MALDI-TOF mass spectrum revealed a moderate content of the cyclic polymers, increasing in parallel to the molecular weights. However, the mass peaks of OH- and Cl-terminated linear chains were detectable, indicating that conversion was far from completion [186]. When more reac-

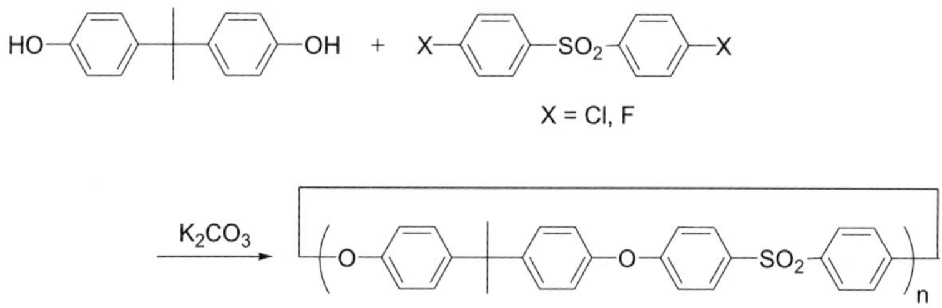

Scheme 48

tive 4,4′-difluorodiphenylsulfone was used instead of the dichloro counterpart and the polymerization conditions were optimized, a polymer with the M_n of 45 000 was obtained. The MALDI-TOF mass spectrum displayed mass peaks of cyclic polymers up to 13 000 Da, and no linear species was detectable [186].

Although rigid-rod poly(*p*-phenyleneterephthalamide) analogues having alkyl side chains did not contain cyclic polymers, the polycondensation of silylated *m*-phenylenediamine and aliphatic dicarboxylic acid chloride afforded cyclic polyamides predominantly (Scheme 49) [187]. Furthermore, cyclic polymers were also produced in polycondensations for polyesters, poly(ether ketone)s, polyimides, and polyurethanes [183]. These examples are the products in polycondensation of AB monomers or in "$A_2 + B_2$" polycondensations, but cyclization of oligomer and polymer was also confirmed in polycondensation of AB_2 monomers [188–195] and in "$A_2 + B_3$" [196–202] and "$A_2 + B_4$" polycondensations [203–206], which afford hyperbranched polymers.

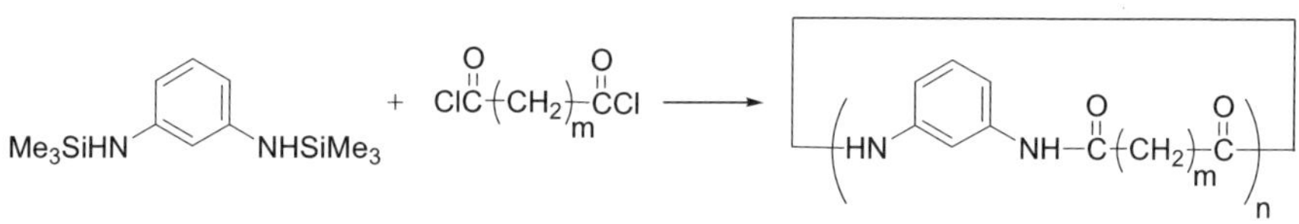

Scheme 49

5.2 Interconversion of Cyclic Oligomer and Chain Polymer

Ring-opening polymerization of cyclic oligomers that are synthesized from polycondensation of monomers has received much attention since Brunelle et al. reported the polymerization of bisphenol A cyclic oligomeric carbonates [207, 208]. The attractiveness of using cyclic oligomers to prepare polymer is that the low molecular weight oligomeric materials have significantly lower viscosities than high molecular weight polymer, thus facilitating the reactive process. Another advantage of using cyclic oligomers for reactive processing is that no volatile or nonvolatile by-products are liberated during polymerization, which resulted in high molecular weight polymer. Brunelle et al. polymerized a mixture of cyclic oligomeric carbonates (dimer to henicosamer) with a catalyst to obtain polycarbonate with high molecular weight (M_w = 50 000–300 000). Nagahata and Sugiyama et al. isolated cyclic dimer to tetramer and studied the polymerization of each cyclic oligomer (Scheme 50). The cyclic dimer has the highest reactivity of the three cyclic oligomers, and polymerized without any catalyst at 300 °C for 5 min to afford an ultrahigh molecular weight polymer estimated as a M_w of 2 000 000 [209]. Although the cyclic tetramer showed much lower reactivity than the cyclic dimer and tetramer, the polymerization was promoted by addition of the cyclic dimer [209] or bisphenol A [210] to afford a high molecular weight polymer (M_w > 1 000 000). An interesting thing in this polymerization is that it proceeded below their melting points, thus the polymerization took place in the solid phase.

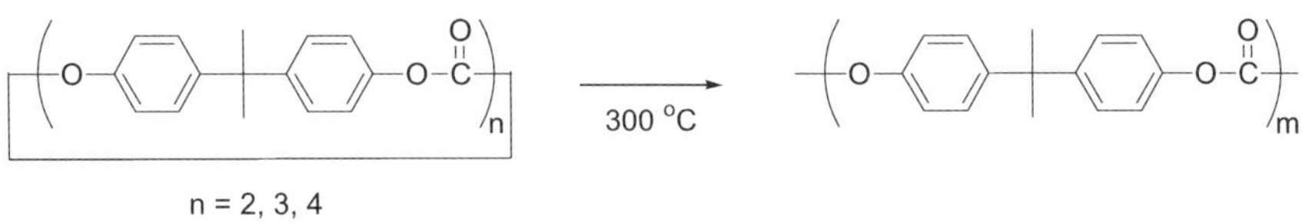

Scheme 50

Brunelle et al. also studied the preparation and ring-opening polymerization of cyclic oligomeric esters [211]. The efficient preparation of alkylene phthalate cyclic oligomers was conducted by the condensation of alkylene diols with iso- and terephthaloyl chlorides in the presence of 2.5–10% of DABCO or quinuclidine and triethylamine making up the remainder of the organic base. The diols and acid chlorides were simultaneously slowly added to a solution of the amines (pseudo-high dilution conditions). The yield of the cyclic oligomers was 49–85%. Ring-opening polymerization using tin or titanate catalysts afforded high molecular weight polyesters within minutes. For example, complete polymerization of PBT cyclic oligomer was able to be achieved at 180–200 °C (the mixture of PBT cyclic oligomers began to melt at about 140 °C and was completely molten at 160–190 °C), significantly below the polymer's melting point of 225 °C, and with molecular weights as high as 445 000 (Scheme 51). Polymers prepared from cyclic oligomers showed higher levels of crystallinity than conventionally prepared polyester. Nagahata et al. reported solid-phase polymerization of ethylene terephthalate cyclic dimer (m.p. = 225 °C) with distannoxane **1** as a catalyst at 200 °C to afford PET with the M_w of 36 100 [212]. They also studied copolymerization of ethylene isophthalate cyclic dimer and bis(2-hydroxyethyl) terephthalate in the presence of a catalytic amount of **1** [213]. Chen and coworkers reported the ring-opening polymerization of phenolphthalein based on polyarylate macrocyclic oligomers [214].

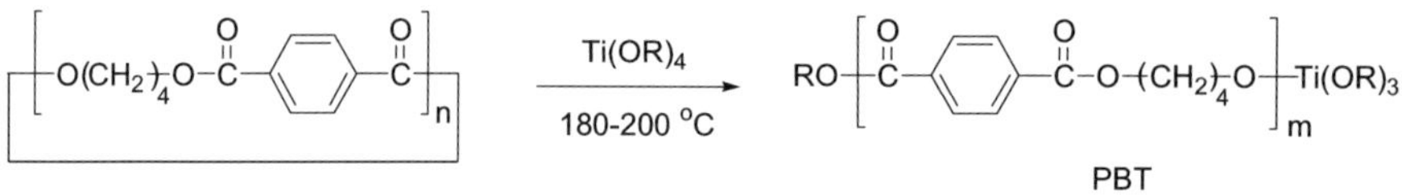

Scheme 51

The cyclic oligomers mentioned above were prepared from polycondensation monomers, whereas another source of considerable interest is from the cyclodepolymerization of the corresponding linear polymer because, when used in combination with the ring-opening polymerization of cyclic oligomers, it provides a possible means of recycling many condensation polymers. Hodge and coworkers investigated the synthesis of cyclic oligo(alkylene isophthalate)s by the cyclodepolymerization of the corresponding linear polymers. For example, poly(ethylene isophthalate) with M_w of 27 000 was treated with 3 mol % of dibutyltin oxide in 1,2-dichlorobenzene (polymer concentration = 2.00% (w/v)) at reflux temperature for 10 days to give the cyclic dimer in 90% yield [215]. Similarly, commercial PEN afforded a mixture of cyclic oligomers in 93% yield. The obtained cyclic oligomers were then heated in the presence of the same tin catalyst at 350 °C for 10 min to give PEN with an inherent viscosity of 0.55 (Scheme 52) [216].

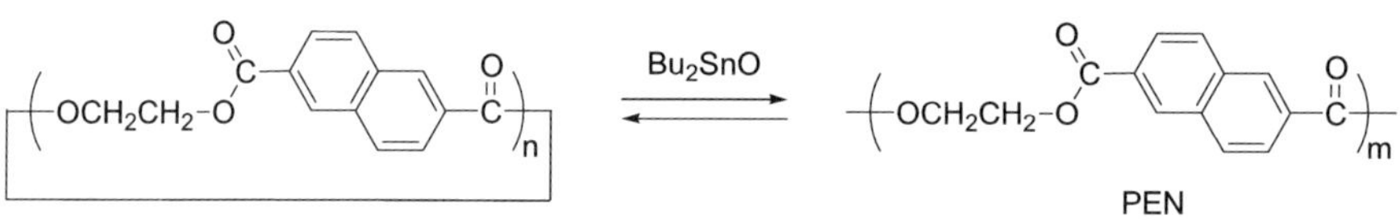

Scheme 52

Regarding poly(ether ketone), cyclic oligomers were prepared from the corresponding monomers under pseudo-high dilution conditions [217–221] and then polymerized with anionic initiators in early works [222–224]. Recently, Colquhoun and coworkers found that catechol-based poly(ether ether ketone) underwent cyclodepolymerization with CsF in DMAc at 230 °C to yield macrocyclic oligomers. The ring-opening polymerization of the macrocyclic oligomers with the cesium salt of 4-hydroxybenzophenone as a catalyst at 350 °C for 15 min regenerated the poly(ether ether ketone) with M_w of 640 000 (Scheme 53) [225]. Cyclic oligomeric ether sulfones were also prepared by fluoride-catalyzed cyclodepolymerization of poly(ether sulfone)s in DMAc at 165 °C [186, 226, 227]. These macrocyclic oligomers underwent efficient ring-opening polymerization in the presence of phenoxide and especially thiophenoxide initiators at 320 °C to regenerate high molecular weight polymer (M_w = 151 000) (Scheme 54) [228]. The cyclic ether sulfone also underwent copolymerization with bisphenoxide and bis(dichlorophenyl)sulfone to give a copolymer of poly(ether sulfone) with

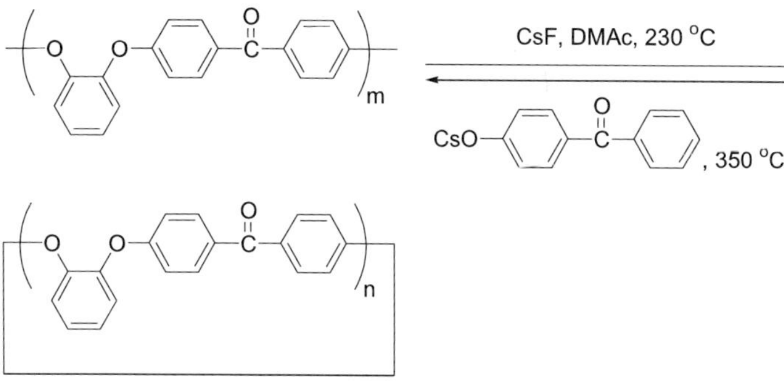

Scheme 53

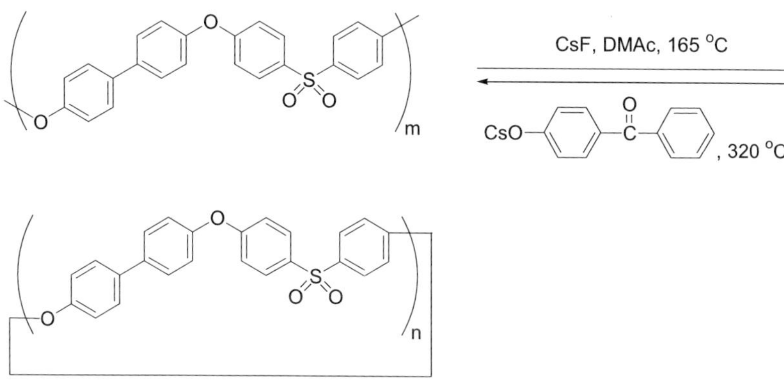

Scheme 54

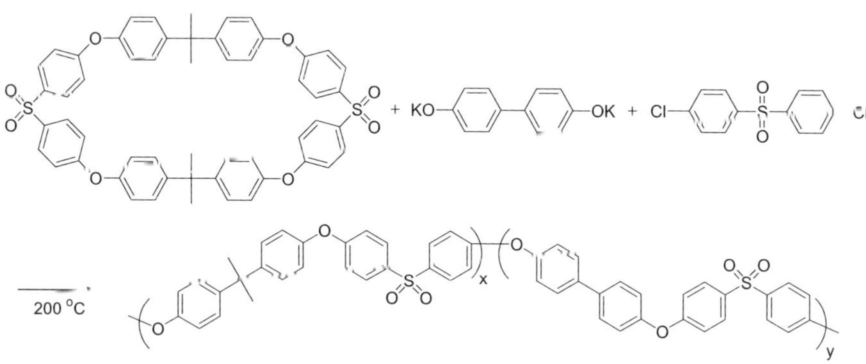

Scheme 55

M_W of 73 000 (Scheme 55) [229]. Colquhoun and coworkers also reported a ring-chain interconversion system in poly(*m*-phenylene isophthalamide) (Nomex) [230]. Cyclodepolymerization was best achieved by treating a solution of the polymer in DMSO containing $CaCl_2$ or LiCl as a solubilizing salt with 3–4 mol% of NaH or the sodium salt of benzanilide as a catalyst at 150 °C. Treatment of a concentrated solution of the macrocyclic oligomers (25% w/v) with 4 mol% of NaH or the sodium salt of benzanilide in a solution of LiCl in DMSO at 170 °C for 6 h resulted in efficient ring-opening polymerization to give poly(*m*-phenylene isophthalamide) with M_w of 46 000 (Scheme 56). These reports demonstrate that the chemical recovery and recycling of high-performance aromatic polymers is in principle entirely feasible.

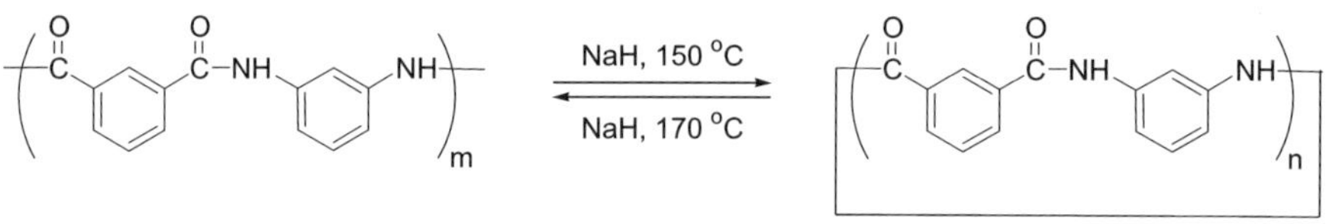

Scheme 56

The macrocyclic aromatic ethers mentioned above are almost strain-free, and their ring-opening polymerization thus depends almost entirely on entropic effects. Colquhoun et al. synthesized highly strained biaryl-containing macrocyclic aromatic ether ketones and ether sulfones by nickel(0)-catalyzed cyclization of the halogen-terminated oligomers [231–233], and conducted the enthalpy-driven ring-opening polymerization of such macrocycles [234]. For example, slow addition of a solution of a linear precursor containing the dibenzofuran structure in DMAc to a solution of $Ni(PPh_3)_4$, generated in situ from $Ni(Py)_4Cl_2$, PPh_3, and Bu_4NI by reaction with excess powdered zinc, in the same solvent resulted in the biaryl macrocycle in 54% yield. The macrocycle polymerized even in solution with potassium 4,4′-biphenoxide at 200 °C to afford polymer with inherent viscosity of 1.10 (Scheme 57). The strain enthalpy of this macrocycle, from DSC analysis of the exothermic ring-opening polymerization was 81 kJ mol^{-1}.

Macrocycles containing the thioether linkage underwent not only anionic ring-opening polymerization [235] but also free-radical ring-opening polymerization [236–248]. For example, Hay and coworkers reported that macrocyclic aryl ether thioether ketone oligomers polymerized in the presence of a catalytic amount of elemental sulfur or 2,2′-dithiobis(benzothiazole) in the melt or in solution at 380 °C to yield high molecular weight polymers (M_w = 143 000) with a small amount of the cyclics remaining (Scheme 58). The free-radical nature of the polymerization reaction was indicated by EPR [238]. Zolotukhin, Colquhoun, and coworkers conducted the ring-opening polymerization of macrocyclic arylene thioether ketones under dynamic heating conditions up to 480 °C. The polymer obtained did not contain residual macrocycles, and the inherent viscosity showed 1.06 (Scheme 59) [248].

These sulfur-containing macrocycles had been prepared under pseudo-high dilution conditions, but Hay and coworkers demonstrated that the cyclodepolymerization of aromatic disulfide polymers was an efficient and easier

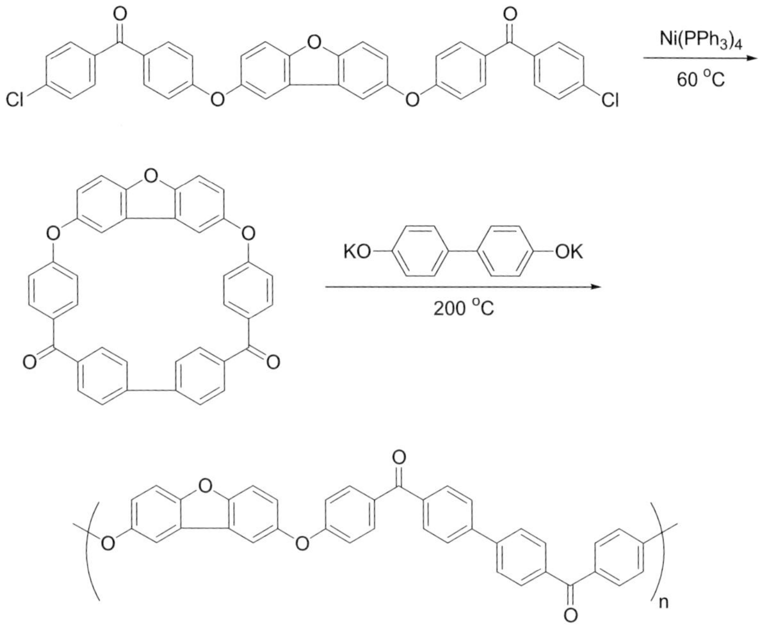

Scheme 57

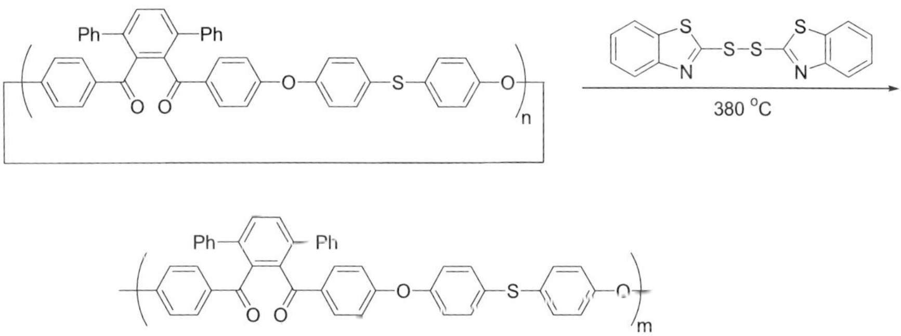

Scheme 58

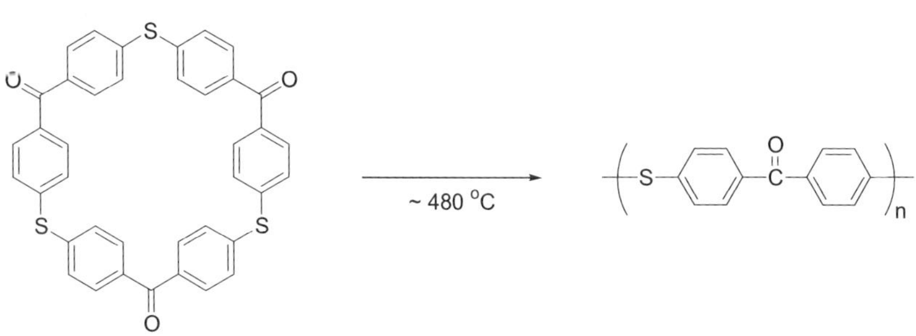

Scheme 59

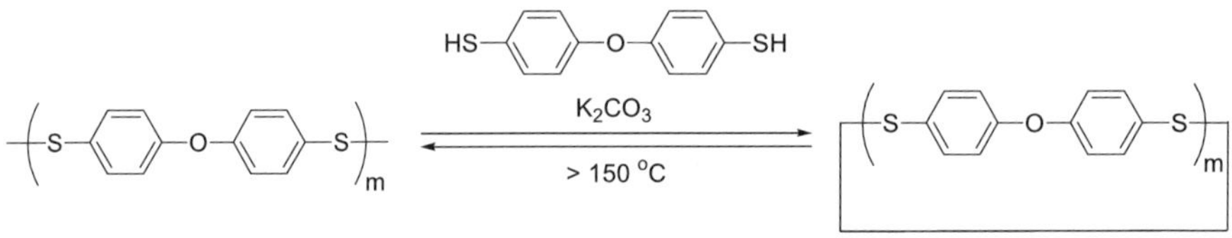

Scheme 60

way to synthesize the corresponding macrocyclic disulfide oligomers [249]. Thus, the disulfide polymer was depolymerized in DMAc at 100 °C for several hours in the presence of a catalytic amount of K_2CO_3 and 4,4′-oxybis(benzenethiol) to afford the macrocycles in more than 98% yield. Melt ring-opening polymerization of the macrocyclic oligomers was performed at 180 °C for 30 min in a nitrogen protective atmosphere. The resulting polymer was insoluble in both DMAc and THF and behaved as an elastic material when heated to the glass transition temperature (Scheme 60) [249, 250].

6 Control of Nonstoichiometric Condensation Polymerization

In condensation polymerization of AA and BB monomers, it is necessary to use exactly equimolar amount of the monomers for production of high molecular weight polymers. However, this fact is based on an assumption that the reactivity of the functional group of monomer is not changed during polymerization. If a functional group in the monomer becomes more reactive when the other functional group has reacted, the condensation polymerization affords a polymer with high molecular weight even when this monomer is used in excess. This is because the AA monomer having the reactivity-enhanced functional groups is liable to react with 2 equivalents of BB to afford BB–AA–BB even though an excess amount of AA is in the reaction mixture. If there is no enhancement of reactivity in the excess monomer AA, both end groups of the oligomer or polymer would be the AA units to terminate the polymerization when all of BB is consumed. This stoichiometrically imbalanced condensation polymerization is desirable, considering that it is sometimes difficult to maintain strict stoichiometric balance between two monomers due to side reactions and physical loss from the reaction medium by evaporation of the monomers or precipitation of the polymer segments.

6.1 Polycondensation of α,α-Dihalogenated Monomers

A well-known example of stoichiometrically imbalanced polycondensation is the polycondensation of dihalomethane and bisphenol derivatives, leading to polyformal (Scheme 61) [251]. In this polycondensation, a polymer with high molecular weight is obtained even when an excess of dihalomethane is used as a solvent. This is attributed to involvement of an active intermediate formed by dihalomethane and the phenol moiety of the bisphenol, which is more reactive than dihalomethane and reacts faster with another phenol moiety. Thus, the polymer end groups are always the phenol moieties, not the halomethyl group, even in the presence of excess dihalomethane, and then polymerization continues without termination with the excess monomer.

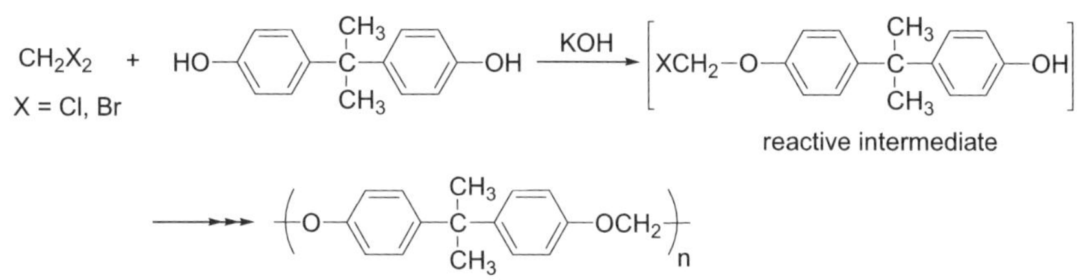

Scheme 61

A practical problem in this polycondensation is formation of cyclic oligomers, resulting in a decrease of the average molecular weight of the polymer [252]. This side reaction is caused by slow dissolution of the potassium salt of bisphenol, which creates high dilution conditions in the reaction mixture. For example, the polymerization was carried out at room temperature by addition of the potassium salt of bisphenol A to a solution of dibromomethane in NMP over several hours to give polyformal, which had a bimodal distribution ($M_n = 46\,900$, $M_w/M_n = 3.1$). In addition, this polyformal contained 23% macrocycles. On the other hand, when bisphenol A and KOH were vigorously stirred in NMP for 2 min, followed by addition of dibromomethane, high molecular weight polymer ($M_n = 250\,100$, $M_w/M_n = 1.3$) was obtained after vigorous stirring for only 5 min. Only miniscule amounts of cyclic oligomers were present. The production of high molecular weight polyformal by high-intensity mixing can be explained as follows (Scheme 62).

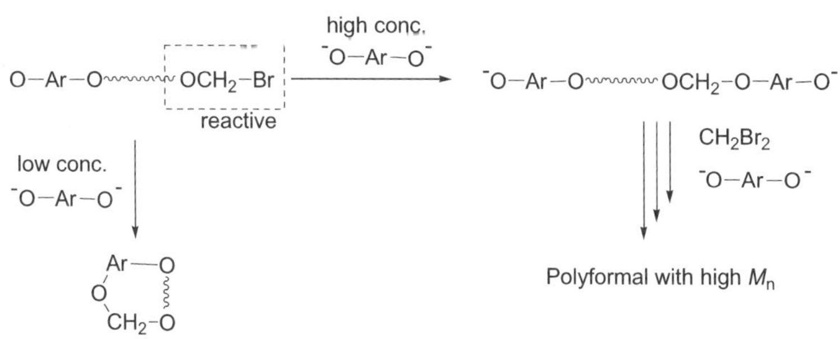

Scheme 62

Since an intermediate bromomethyl ether is much more reactive than dibromomethane, as mentioned above, the intermediate could react with itself to give a cycle under normal stirring conditions where the salt of bisphenol very slowly goes into the solution. Under high-intensity mixing conditions the concentration of the salt of bisphenol is high, so that the intermediate would preferentially react with the salt to give high molecular weight polymer while suppressing the formation of macrocycles.

The polycondensation of dibromomethane with bis(benzenethiol) instead of bisphenol also shows stoichiometrically imbalanced polymerization behavior (Scheme 63) [253]. When 1.5 equivalents of dibromomethane was reacted with dithiol in NMP at 75 °C for 4 h, polysulfide with the maximum inherent viscosity ($\eta_{inh} = 0.50\ dL\,g^{-1}$) was obtained. On the basis of the model reaction, a reactive intermediate, in which a bromine in dibromomethane is substituted with the mercapto group of the monomer, is estimated to be 61 times more reactive than dibromomethane.

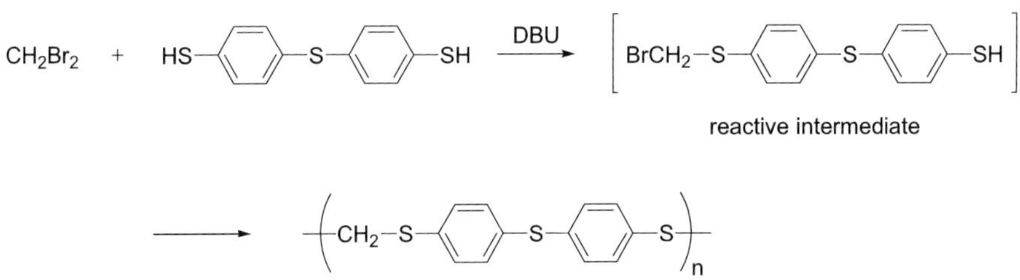

Scheme 63

Not only dihalomethane but also 2,2-dichloro-1,3-benzodioxole (**12**) affords a reactive intermediate by the reaction with bisphenol A (Scheme 64) [254]. Thus, the polycondensation of 1.7 equivalents of **12** with bisphenol A in refluxing dichlorobenzene for 3 h yielded polyorthocarbonate with the highest

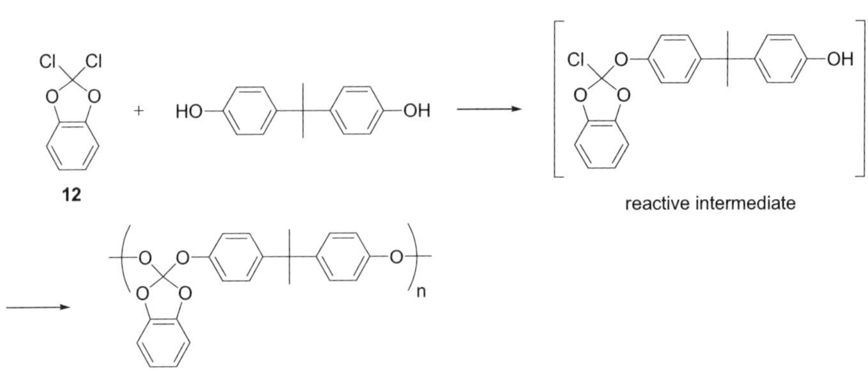

Scheme 64

molecular weight (M_n = 120 000). In the model reaction of **12** and phenol at −40 °C, the rate of the second nucleophilic substitution of monochloride formed by the first substitution was 27 times faster than that of the first substitution reaction.

6.2 Palladium-Catalyzed Polycondensation

The above nonstoichiometric polycondensations are derived from the specific reactivity of electrophilic monomers. Catalysts for polycondensation also bring about these polycondensations without strict stoichiometric balance between the monomers. Nomura et al. applied palladium-catalyzed allylic substitution (Tsuji–Trost reaction) to the polycondensation between 1,4-diacetoxybut-2-ene (**13**) and diethyl malonate in the presence of a Pd(0) catalyst, resulting in high molecular weight polymer (M_n = 21 800) (Scheme 65) [255, 256]. Surprisingly, the polycondensation by using excess of **13** was also found to afford similar high molecular weight polymers [257].

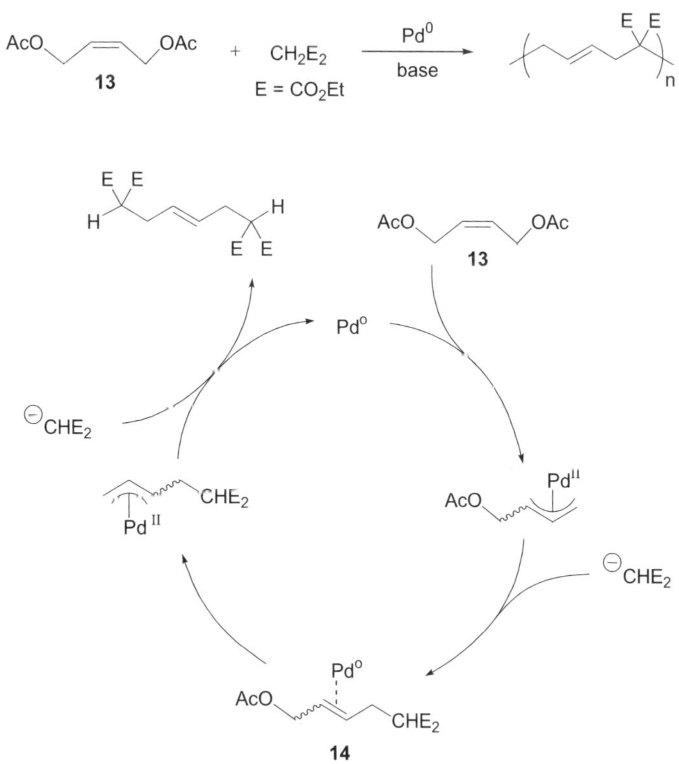

Scheme 65

The observed stoichiometrically imbalanced polymerization behavior is rationalized by involvement of olefin-Pd(0) complex **14**, which leads to cascade bidirectional allylation. Thus, after the first allylation of **13**, **14** selectively forms the allylpalladium(II) complex at the other allylic terminal. Therefore, the polymer end groups are always the malonic ester moiety even in the presence of excess **13**, which would not terminate the polycondensation by the attack on both polymer end groups. This behavior is dependent on the ligand of the Pd(0) catalyst; bis(diphenylphosphino)butane (dppb) is indispensable, and use of PPh_3 results in low molecular weight polymer as expected by Carothers and Flory's basic principle [1–3].

The Pd-catalyzed nonstoichiometric polycondensations of other monomers, in which an olefin-Pd(0) complex after the first allylic substitution exchanges the olefin ligand with the remote double bond, was also reported [258]. The dependence of nonstoichiometric polycondensation behavior on a catalyst is interesting in comparison to the polycondensation mentioned above, where that behavior only stems from the structure of the monomers.

6.3 Crystallization Polycondensation

When polycondensation of AB monomer is carried out in the presence of a monofunctional compound, polycondensation is terminated with the monofunctional compound, and the DP is estimated by Carothers and Flory's basic principle. However, Kimura et al. reported that the polycondensation of 4-acetoxybenzoic acid in the presence of monofunctional compound in liquid paraffin at 320 °C affords poly(4-oxybenzoyl) crystal (whisker) consisting of higher molecular weight polymers compared with the products obtained by melt polymerization (Scheme 66) [259]. The DP of the aromatic polyester obtained was much higher than that calculated according to the Carothers and Flory equation. For example, in the polymerization with 30 mol % of the monofunctional compound, where the calculated DP is 3.3, the observed DPs were 232 ($R = CH_3(CH_2)_9$) and 283 ($R = CH_3(CH_2)_{17}$). This polymerization would proceed through the following mechanisms. Oligomers are formed in the solution, and when the DP of the oligomers exceeds a critical value, they are crystallized. End-free oligomers are preferentially crystallized due to a solubility lower than that of end-capped oligomers. End-capped oligomers are also crystallized and polycondensation proceeds with elimination of the end-

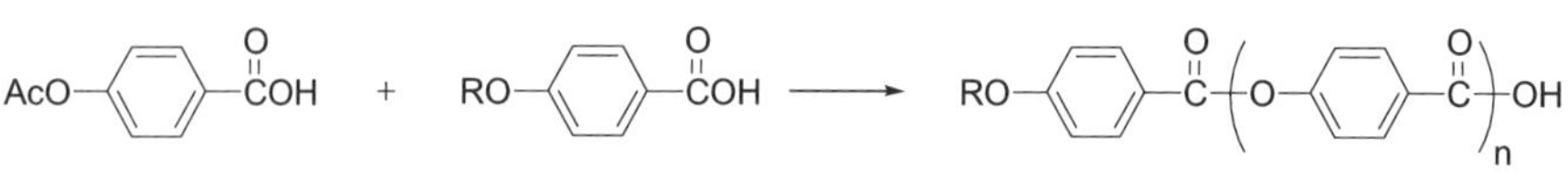

Scheme 66

capping groups of the oligomers by transesterification just when they are crystallized. Although some of the end-capped oligomers are contained in the crystals, the end-capping groups are excluded by solid-state polycondensation.

Similar polycondensation giving crystals was attained by the reaction of 4-hydroxybenzoic acid with 4-ethoxybenzoic acid anhydride, which immediately affords 4-(4-ethoxybenzoyloxy)benzoic acid, a monomer, and monofunctional compound [260]. Polyesters with a DP of 38–76 were obtained even if 30–60 mol% of the anhydride was added to the reaction mixture (Scheme 67). This polycondensation is noteworthy not only as a stoichiometrically imbalanced polycondensation, but also as a valuable method for morphology control of condensation polymers during polymerization [261].

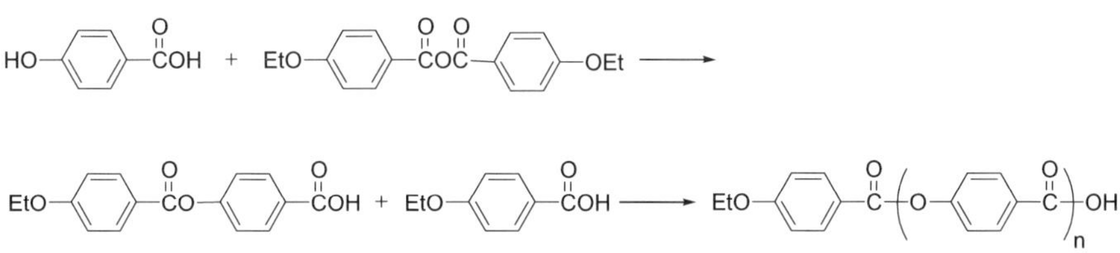

Scheme 67

6.4 Nucleation–Elongation Polycondensation

The closed-system, reversible polymerization of *m*-phenylene ethynylene starter sequences **15** and **16** in solution was driven by the folding energy of the resulting polymers [262, 263]. Zhao and Moore also performed this polymerization under conditions of imbalanced stoichiometry to obtain polymers whose molecular weight was higher than expected from the Flory distribution [264] (Scheme 68). For instance, in the reaction of **15** and **16** at a molar ratio of 1:2 in the presence of oxalic acid at room temperature, polymer products were present at equilibrium along with a substantial amount of monomer but relatively little dimer or trimer. When the polymerization of the corresponding diamine and dialdehyde at the same monomer stoichiometry was carried out in the melt, only low molecular weight oligomers were produced, and the DP was in good agreement with the prediction from the Flory equation. Because the helical folding of *m*-phenylene ethynylene chains should not take place under these melt conditions, the stoichiometrically imbalanced polymerization of **15** and **16** in solution was proposed to proceed in the folding-driven nucleation–elongation mechanism. Thus, an oligomer with a certain length starts folding in solution, and further extending the molecule would result in polymers with increasing folding stability. Excess monomer remains unreacted at equilibrium because it would not take part in producing the folded polymer.

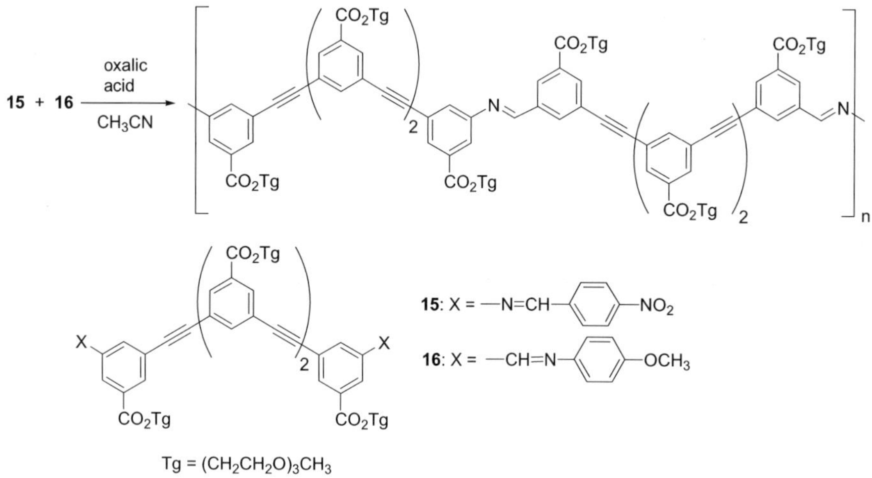

Scheme 68

7 Control of Molecular Weight and Polydispersity

Until quite recently, controlling molecular weight and polydispersity of a polymer in condensation polymerization was thought to be essentially impossible, because condensation polymerization proceeds in a step-growth polymerization manner: the polymerization is initiated by the reaction of monomers with each other, and propagation involves the reaction between both end groups of all types of oligomers that are formed, as well as the reaction of the oligomers with the monomer. In contrast, nature synthesizes monodispersed biopolymers, such as DNA and polypeptides, by condensation polymerization. This system involves selective reaction of monomers with the polymer end group that is activated by an enzyme, which is chain-growth polymerization.

Accordingly, artificial condensation polymerization is also able to control molecular weight and polydispersity when the condensation polymerization proceeds in a chain-growth polymerization manner similar to biological polycondensation (we call it chain-growth condensation polymerization). For chain-growth condensation polymerization, it is necessary that the polymer end group is always more reactive than the monomer to prevent the monomer from reacting with another monomer, leading to step-growth polymerization. There are four approaches to maintain the reactive polymer end group during polymerization: the first is transfer of reactive species, derived from the initiator, to the polymer end group; the second is activation of the polymer end group by different substituent effects between the monomer and polymer; the third is transfer of the catalyst to the polymer end group like biological condensation polymerization; and the fourth is biphasic polymerization, in which solid monomer, transferred with PTC to a solution of polymer, reacts with the polymer end group.

7.1 Transfer of Reactive Species

Specific elements can transmit a reactive species, stemmed from an initiator, with elimination of small molecules. The first category is insertion of the monomer to the terminal M–F bond (Scheme 69). The second one is transfer of the cationic species (Scheme 70).

Scheme 69

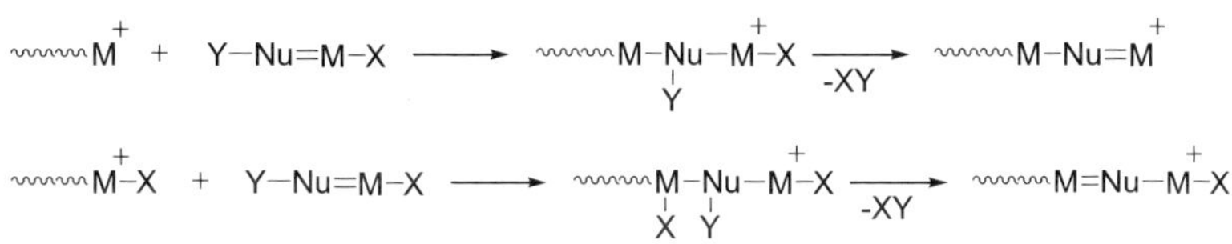

Scheme 70

The polymerization of dimethylsulfoxonium methylide (**17**) initiated by trialkylborane, which Shea et al. developed, is classified as the first category. Propagation involves insertion of **17** into the terminal C–B bond with elimination of DMSO. The polymerization was carried out in toluene at 70–80 °C, followed by oxidative workup to yield hydroxyl-terminated polymethylene (Scheme 71). The M_n values were very close to the calculated values from the feed ratio of **17** to trialkylborane, and the M_w/M_n ratio ranged from 1.04 to 1.17. These results are consistent with the character of living polymerization.

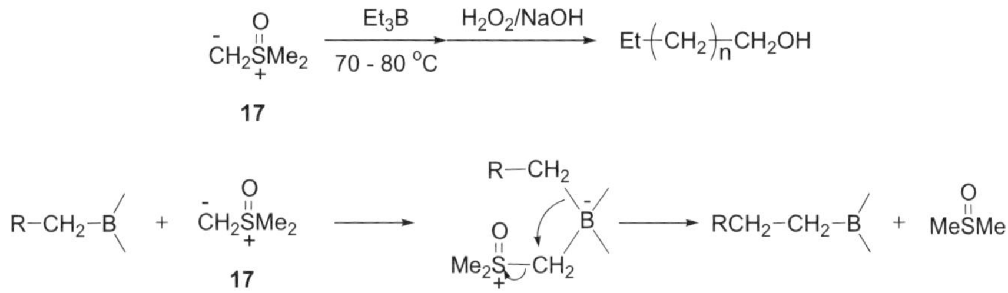

Scheme 71

The insertion mechanism involves initial attack of the ylide on the alkylborane. The borate complex undergoes 1,2-migration of the alkyl group to produce the homologated alkylborane and a molecule of DMSO [265, 266].

This polyhomologation with alkylboranes is amenable to the synthesis of telechelic polymethylene, because alkylboranes can be prepared by hydroboration of a variety of α-olefins. For example, polymethylene with 4-methoxyphenyl and hydroxy groups at both ends was synthesized from an initiator prepared by hydroboration of 4-vinylanisole with $BH_3{\cdot}THF$ (Scheme 72) [265]. Other functional groups including biotin, carbohydrates, primary and secondary amines, and dansyl and pyrene fluorescent groups were also introduced as an end group from the corresponding α-olefins [267]. When *B*-thexylboracycloheptane, prepared by the hydroboration of 1,5-heptadiene with thexylborane, was used as an initiator, methylene was inserted into only the C–B bond of boracyclane, not the thexyl–B bond, resulting in ring expansion. The expanded boracycle was treated with sodium cyanide, followed by benzoyl chloride then peroxide oxidation to yield cyclic ketone (Scheme 73) [268]. Furthermore, tri-armed star polymethylenes were also synthesized by this polymerization method [269, 270].

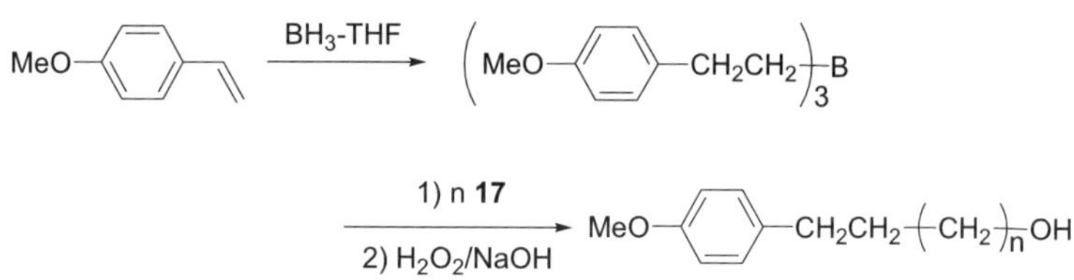

Scheme 72

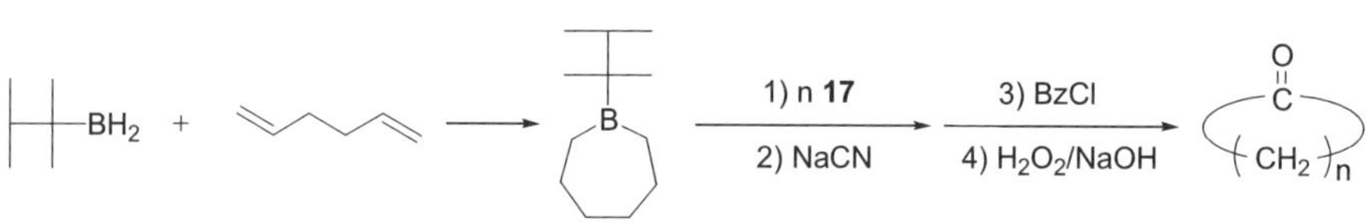

Scheme 73

Mioskowski and coworkers reported similar polymerization of arsonium ylide **18** initiated by trialkylborane [271, 272]. However, the reaction of **18** at 0 °C in THF did not lead to polymers substituted on every carbon atom like the polymerization of sulfoxonium ylide **17**, but rather to a polymer in which the main chain was elongated by three carbon atoms at a time (Scheme 74). Polymers of varying DPs were obtained by using varying ratios of ylide **18** to trialkylborane, but the DPs were larger than expected from these ratios. The M_w/M_n ratio ranged from 1.21 to 1.58. This suggests that the initiation of the polymerization was not completely efficient, but the propagation was

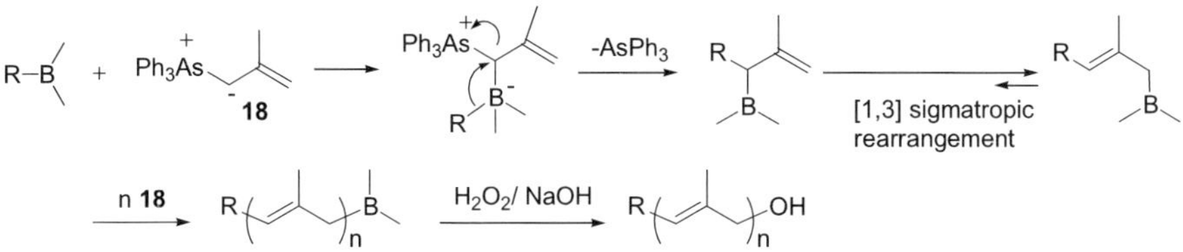

Scheme 74

controlled. The polymerization mechanism, which is different from that of the polymerization of ylide **17**, involves [1,3] sigmatropic rearrangement after 1,2-migration of the alkyl group of borate. This is the reason for the elongation of three carbons in this insertion polymerization. Block copolymers were also synthesized by successive polymerization of different arsonium ylides (Scheme 75) [272].

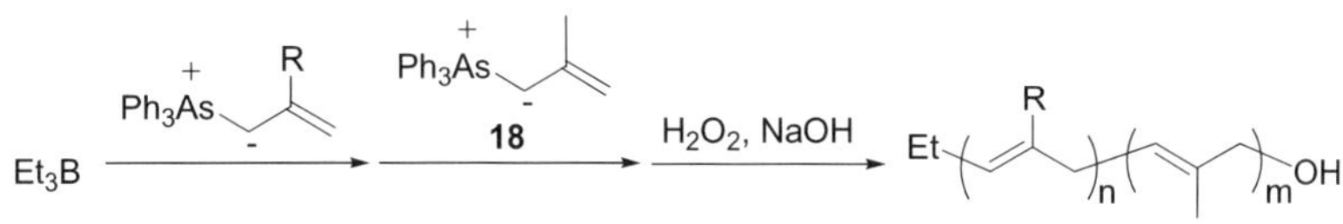

Scheme 75

An example of the second category is the polymerization of phosphoranimines **19** initiated with electrophiles, leading to polyphosphazenes. Matyjaszewski and coworkers found that *N*-silylated phosphoranimine **19a** polymerized with $SbCl_5$ at 100 °C to yield poly[bis(trifluoroethoxy)phosphazene] **20** with M_n of 10 000–50 000 and M_w/M_n of 1.2–2.5 [273]. The molecular weight reached a maximum at partial conversion and leveled off, most likely due to transfer reactions. However, the inverse relationship of molecular weight and $SbCl_5$, the appearance of high molecular weight polymer at partial conversion, and the first order in $SbCl_5$ kinetics of the reaction supported the notion that $SbCl_5$ was the true initiator and that polymerization proceeded in a chain-growth manner. The zero order in monomer kinetics of the reaction indicates that a reactive intermediate is formed in fast equilibrium with the monomer, followed by slow unimolecular elimination of trimethylsilyl trifluoroethoxide (Scheme 76).

Allcock and coworkers reported that trichloro(trimethylsilyl)phosphoranimine **19b** polymerized with PCl_5 at ambient temperature with elimination of trimethylsilyl chloride. The resultant poly(dichlorophosphazene) was treated with an excess of $NaOCH_2CF_3$ to give polymer **20**. When the polymerization was carried out in dichloromethane, the molecular weight increased with increase in the ratio of **19b** to PCl_5 ($M_w = 7000–14\,000$), and the molecular weight distribution was kept narrow ($M_w/M_n = 1.04–1.20$) [274]. The polymerization in toluene proceeded faster than in dichloromethane to give

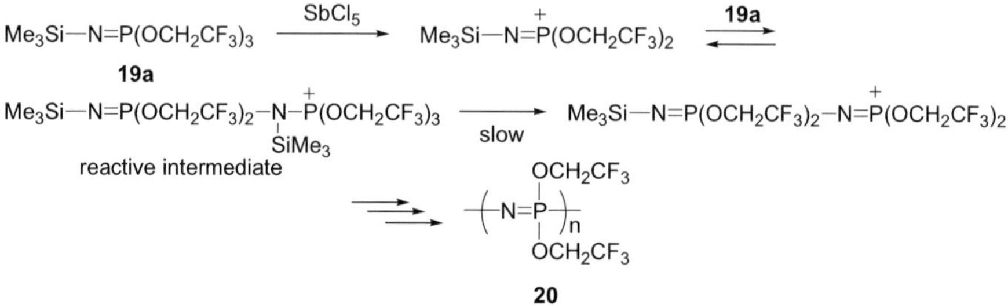

Scheme 76

polymers with higher molecular weights in the range of 10^5 with low polydispersities (<1.3) [275]. Phenyl-substituted monomer **19c**, $PhCl_2P = NSiMe_3$, also underwent controlled polymerization until the feed ratio of **19c** to PCl_5 was 100. Other related initiators such as $SbCl_5$, $TaCl_5$, or $PhPCl_4$ also appear to initiate the ambient temperature polymerization of **19b,c**. The polymerization is initiated by the reaction of **19b** with 2 equivalents of PCl_5 with elimination of Me_3SiCl to generate a salt, to which **19b** successively reacts with elimination of Me_3SiCl, resulting in the elongated cation (Scheme 77).

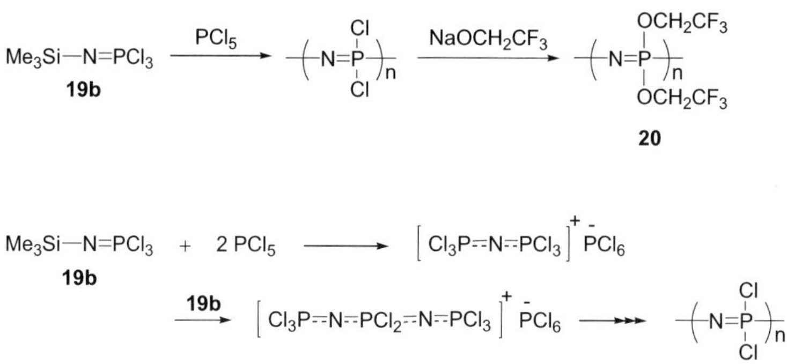

Scheme 77

Monomer **19b** was synthesized by the reaction of PCl_5 with either $LiN(SiMe_3)_2$ or $N(SiMe_3)_3$. These methods give relatively low product yields, because PCl_5 is an initiator for the polymerization of **19b**. To circumvent this concurrent polymerization, a new method for synthesizing **19b** and the subsequent polymerization in one pot has been recently reported [276]. Thus, PCl_3 was reacted with $LiN(SiMe_3)_2$ to afford $Cl_2P-N(SiMe_3)_2$, which was oxidized with SO_2Cl_2 to yield **19b**. Into the mixture, mainly containing **19b**, Me_3SiCl, and LiCl, was added PCl_5 to produce the polyphosphazene. Under the optimized conditions, the derivative polymer **20** showed a relatively narrow molecular weight distribution ($M_w/M_n = 1.24$) even in one-pot reaction from PCl_3 (Scheme 78). Monomer **19a** also polymerized with an anionic catalyst via a chain-growth process, although this polymerization is not included

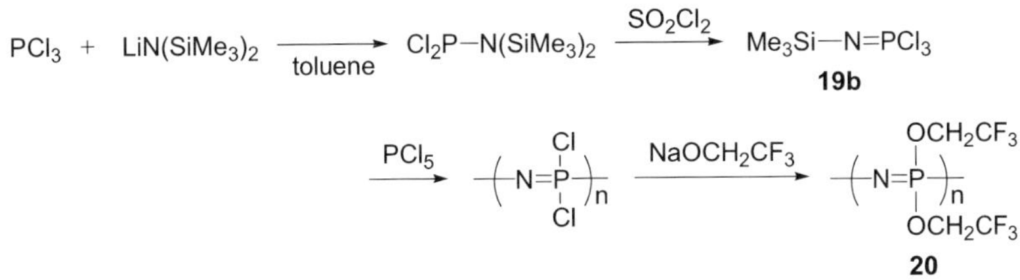

Scheme 78

in the second category of Scheme 78. Montague and Matyjaszewski reported this polymerization before the cationic polymerization mentioned above. The polymerization was carried out with TBAF as an initiator at 95 °C to yield **20** with M_w/M_n of about 1.5 [277]. The proposed polymerization mechanism is as follows. The polymerization is initiated by the abstraction of the silyl group from **19a** with TBAF, followed by the attack of the resultant phosphazene anion on another monomer. Propagation proceeds through the attack of the resultant trifluoroethoxide on the silyl group of the growing polymer chain, producing an anion which can then attack another monomer molecule (Scheme 79). Selective desilylation of the polymer end group can be explained by the possibility that the strength of the N–Si bond on a polymer end group is weaker than that on a monomer molecule due to long conjugation of the polymer [278].

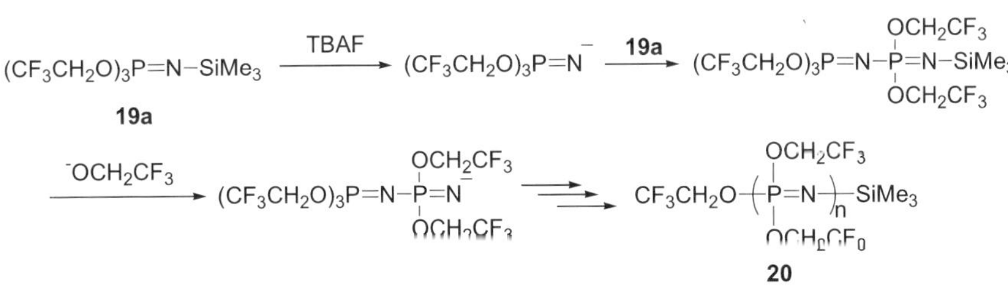

Scheme 79

Polyphosphazene block copolymers were synthesized by these chain-growth polymerization methods. The successive anionic polymerization of *N*-silylphosphoranimines **19d** and **19a** at 133 °C yielded the block copolymer with M_w/M_n of 1.4–2.3 (Scheme 80) [278, 279]. However, due to the presence of two possible leaving groups in **19d**, this approach yielded block copolymers where one of the block segments contained a mixture of side groups. On the other hand, the cationic polymerization of **19b** with PCl_5 was carried out at ambient temperature, followed by addition of a second phosphoranimine to yield a block copolymer with M_w/M_n of 1.1–1.4, where each block segment had one kind of side chain (Scheme 81) [280].

Some block copolymers of polyphosphazene and conventional polymers were also synthesized. The first example is a block copolymer of polyphos-

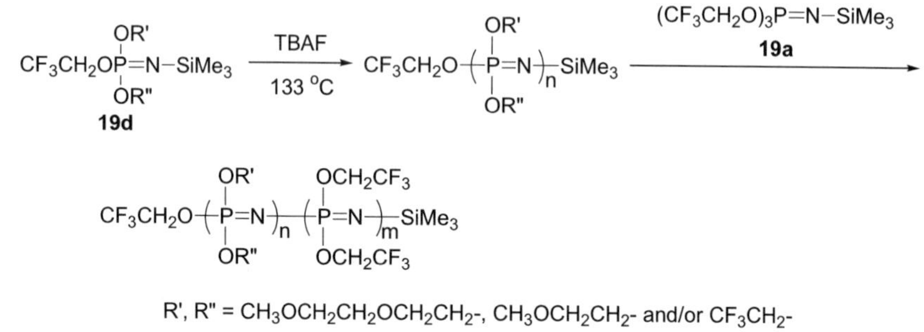

Scheme 80

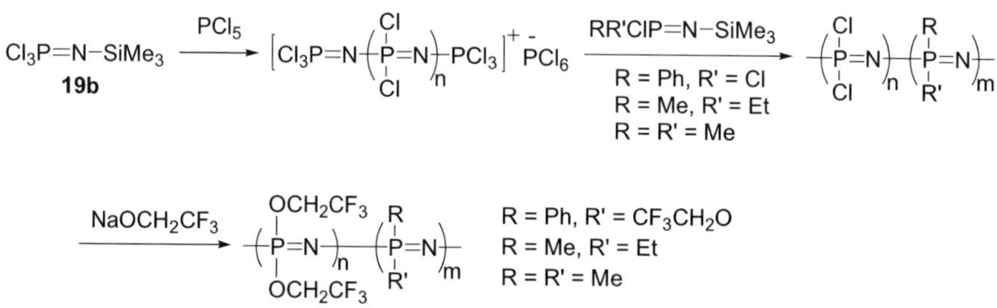

Scheme 81

phazene and PEG. Amino-terminated PEG was reacted with bromophosphoranimine **21** in the presence of triethylamine to afford PEG-phosphoranimine, which was treated with 2 equivalents of PCl_5 at −78 °C, resulting in the formation of a macroinitiator. This macroinitiator induced chain-growth polymerization of **19b** to yield the diblock copolymer. Following termination, the chlorine atoms were replaced with trifluoroethoxide groups (Scheme 82). When difunctional amino-terminated PEG was used, the triblock copolymer was obtained [281, 282]. The micellar characteristics of this amphiphilic di-

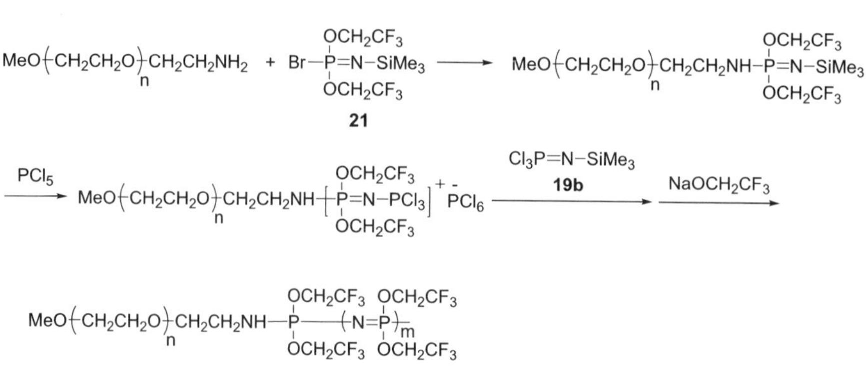

Scheme 82

block copolymer were studied [283]. Triblock copolymers consisting of PEG and two kinds of polyphosphazenes were also synthesized [284].

Di- and triblock copolymers of polyphosphazene and polystyrene were prepared by the macroterminator method. Phosphine-terminated polystyrene, prepared by quenching anionic living polystyrene with Ph_2PCl, was treated with N_3SiMe_3 to afford polystyryl phosphoranimine. This species was used as a macromolecular terminator for monofunctional and difunctional living poly(dichlorophosphazene)s, derived from the cationic polymerization of **19b**, to yield the di- and triblock copolymers, respectively. Following termination, the chlorine atoms were replaced with trifluoroethoxide groups [285] (Scheme 83). The monofunctional poly(dichlorophosphazene) was prepared by using nonhalogen phosphoranimines, such as $(CF_3CH_2O)_3P = NSiMe_3$, to generate the initiating species with PCl_5 for the living polymerization of **19b** [286]. Diblock copolymers of polystyrene and polyphosphazene with a diethylene glycol monomethyl ether side chain were also synthesized, and the self-association behavior of the block copolymers in aqueous media was investigated [287].

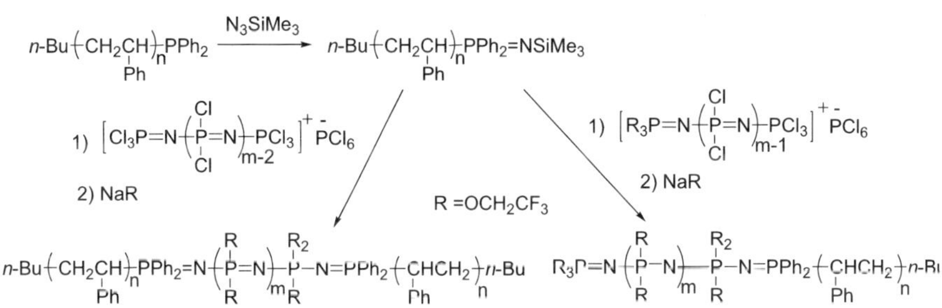

Scheme 83

Since the procedure for the formation of macroinitiator by using **21**, mentioned in the synthesis of block copolymers of polyphosphazene and PEG, enabled the synthesis of monofunctional polyphosphazenes [288], monoallyl functional polyphosphazene was prepared and underwent hydrosilylation with dihydride-terminated poly(dimethylsiloxane) to pro-

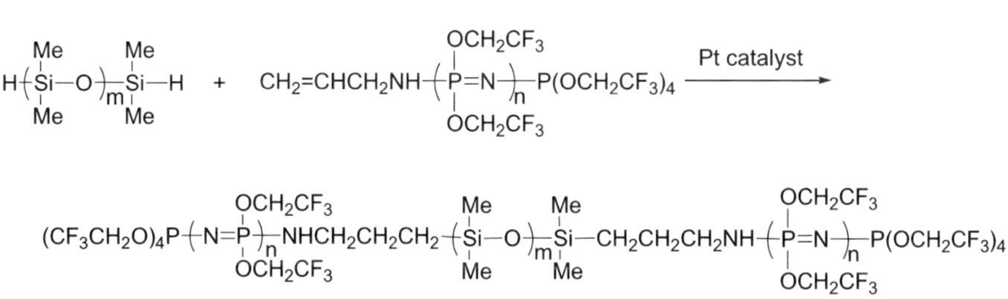

Scheme 84

duce polyphosphazene–polysiloxane–polyphosphazene triblock copolymer (Scheme 84). When phosphoranimine-terminated polysiloxane was used, polysiloxane–polyphosphazene–polysiloxane triblock copolymer was obtained in a similar manner to that mentioned in the synthesis of block copolymers of polyphosphazene and polystyrene [289, 290]. Tri-armed star polyphosphazene [291] and graft copolymers of conventional polymer with polyphosphazene [285, 292, 293] were also reported.

7.2 Different Substituent Effects Between Monomer and Polymer

Even general AB-type monomers, affording polyamide, polyester, polyether, and so on, undergo chain-growth condensation polymerization if the polymer end group becomes more reactive than the monomer by virtue of the change of substituent effects between the monomer and polymer. Both the resonance effect and inductive effect of the nucleophilic site on the reactivity of the electrophilic site at the *para* and *meta* positions of the monomer are applicable, respectively.

7.2.1 Resonance Effect (Polymerization of *p*-Substituted Monomers)

In polycondensation of some *p*-substituted monomers, the polymer end group was reported to become more reactive than the monomer, but the polycondensation did not precisely control the molecular weight of polymer within narrow molecular weight distributions because of the insolubility of the polymers and the contamination of step-growth condensation polymerization [294–299]. However, we found that the polycondensation of phenyl 4-(octylamino)benzoate (**22a**) proceeded homogeneously in the presence of a base (a combination of *N*-octyl-*N*-triethylsilylaniline, CsF, and 18-crown-6) and phenyl 4-nitrobenzoate (**23**) as an initiator in THF at ambient temperature to yield well-defined aromatic polyamides with very low polydispersities ($M_w/M_n \leq 1.1$) [300]. The M_n of the polymer was controlled by the feed ratio of monomer **22a** to initiator **23** up to 22 000, and the polydispersity was quite narrow. Furthermore, the M_n values also increased in proportion to monomer conversion, indicating that this condensation polymerization proceeded in the manner of chain-growth polymerization. This result is explained by the different substituent effects between the monomer and polymer (Scheme 85). The base abstracts the proton of the amino group of monomer **22a** to generate the aminyl anion **22a′**. This anion deactivates the phenyl ester moiety of **22a′** by its strong electron-donating ability through the resonance effect, resulting in preventing the monomer from reacting with another one. The anion monomer **22a′** would react with initiator **23** having an electron-withdrawing group, because the phenyl ester moiety of **23** is more reactive than that of

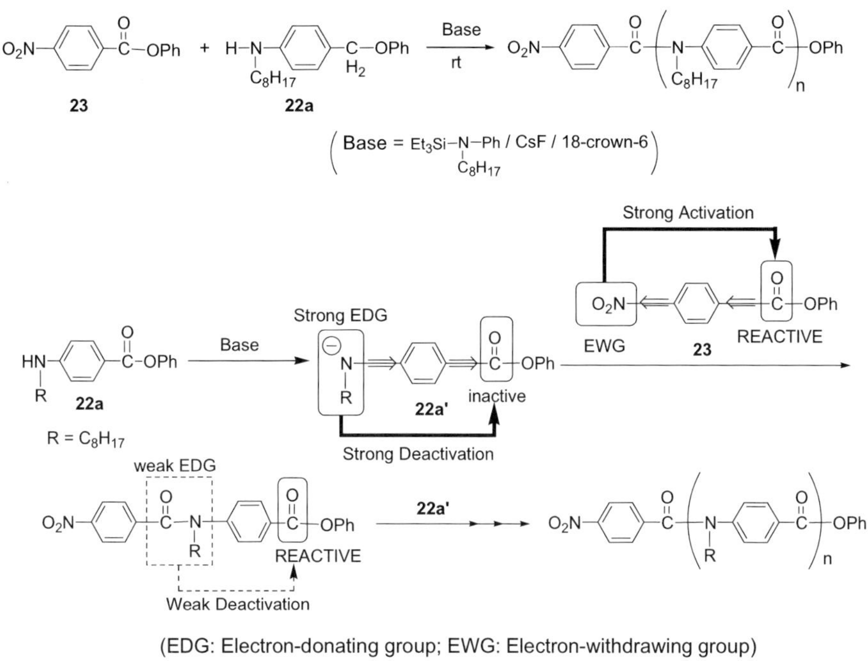

Scheme 85

22a′. The obtained amide has a weak electron-donating amide linkage, and the phenyl ester moiety of the amide is more reactive than that of **22a′**. Thus, the next monomer would selectively react with the phenyl ester moiety of the amide. Growth would continue in a chain polymerization manner by the selective reaction of **22a′** with the polymer terminal phenyl ester moiety

The above synthetic method for well-defined aromatic polyamides, however, needs a peculiar base, *N*-octyl-*N*-triethylsilylaniline, along with CsF and 18-crown-6, and the monomer has a phenyl ester moiety as an electrophilic site, which is not that common compared with a methyl ester or an ethyl ester. Furthermore, it is necessary to separate the obtained polyamide from by-products, such as *N*-octylaniline and phenol, by HPLC. For convenient synthesis, the polycondensation of the corresponding methyl ester monomer **22b** with a commercially available base has been developed [301]. The methyl ester **22b** polymerized with LiHMDS in the presence of an initiator in THF at −10 °C (Scheme 86). The highly pure polyamide with a defined molecu-

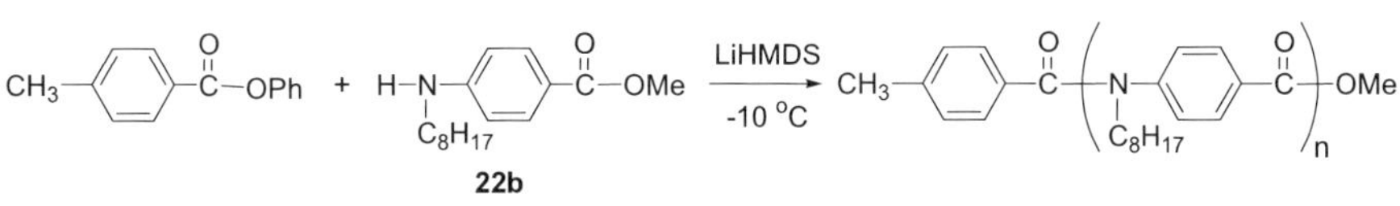

Scheme 86

lar weight and low polydispersity was obtained after simple treatment of the reaction mixture with aqueous NaOH solution followed by evaporation, because the by-products in this condensation polymerization after treatment with water are low-boiling methanol, ammonia, and hexamethyldisiloxane. The condensation polymerization of a similar monomer containing a 3-acyl-2-benzothiazolethione as the electrophilic site instead of the methyl ester moiety of **22b** would also make the purification of the obtained polyamide easier; by-products could be washed out with water [302].

Block copolymers of aromatic polyamides were synthesized by the chain-growth condensation polymerization of a variety of 4-(alkylamino)benzoic acid esters **22**. An example of the block copolymer of *N*-alkyl and *N*-H polyamides is shown in Scheme 87 [303]. First, *N*-octyl monomer **22a** was polymerized, and then monomer **22c** having a protecting group on the amino group and a base were added to the reaction mixture. The added **22c** polymerized smoothly from the end of poly**22a** to yield the block copolymer of poly**22a** and poly**22c**. The protecting group was quantitatively removed with TFA to afford the desired block copolymer of *N*-alkyl and *N*-H polyamides with narrow polydispersity. The reason why **22c** was used for this block copolymerization was that a monomer having a primary amino group did not polymerize under these polymerization conditions [304]. The block copolymer was arranged through self-assembly in THF by virtue of intermolecular hydrogen bonding of the *N*-H polyamide unit. SEM images showed micrometer-sized bundles and aggregates of flake-like structures. Recently, block copolymers of *N*-octyl and *N*-fluoroalkyl polyamides with narrow polydispersity were synthesized and their self-assembly was studied [305, 306].

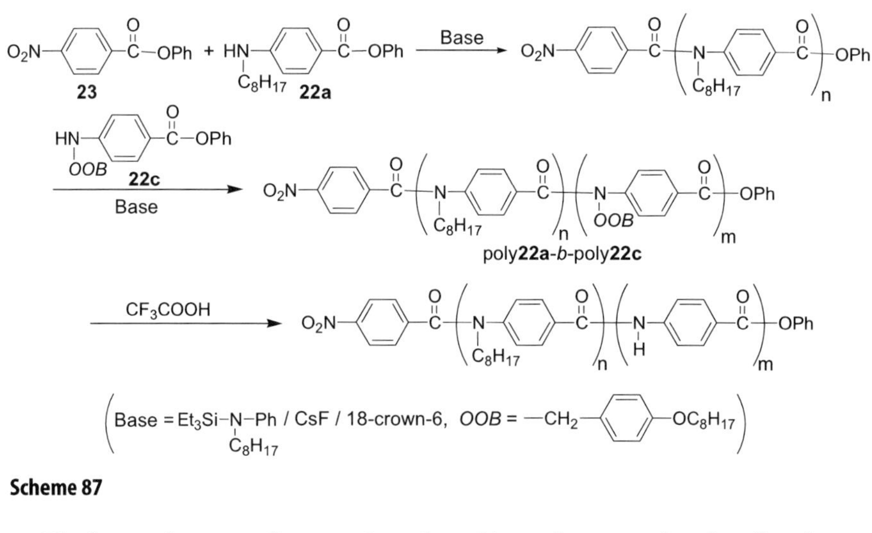

Scheme 87

Block copolymers of aromatic polyamide and conventional coil polymer are prepared by the reaction of the polymer end group of the polyamide with the living propagating group of the coil polymer. Thus, the phenyl ester moiety of the polyamide reacts with the anionic living end of the coil polymer, whereas the amino group of the polyamide reacts with the cationic living end of the coil polymer. For example, PEG monomethyl ether was reacted with the polyamide prepared by the chain-growth polycondensation of **22a** in the presence of NaH to yield a block copolymer of polyamide and PEG (Scheme 88). Excess PEG was used in this polymer reaction, but unreacted PEG could be washed out with water to isolate the block copolymer [304, 307]. Similar reaction of PEG with a polyamide obtained by the chain-growth polycondensation of **22a** with phenyl terephthalate as a bifunctional initiator gave a triblock copolymer of PEG–aromatic polyamide–PEG [308].

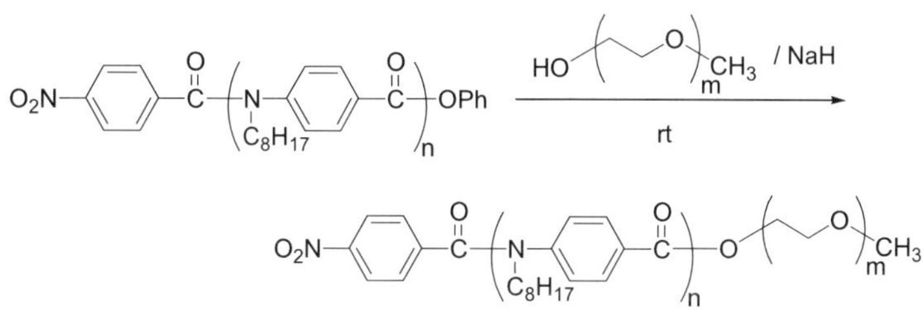

Scheme 88

Polyamide with a terminal amino group was prepared by the polymerization of **22a** with an initiator bearing the BOC group on the amino group, followed by treatment with TFA to remove the BOC group. The terminal amino group of the polymer reacted with living poly(THF) to yield a block copolymer of polyamide and poly(THF) (Scheme 89). When difunctional living poly(THF) initiated by trifluoromethanesulfonic anhydride was reacted with the above polyamide, polyamide–poly(THF)–polyamide triblock copolymer was produced [309].

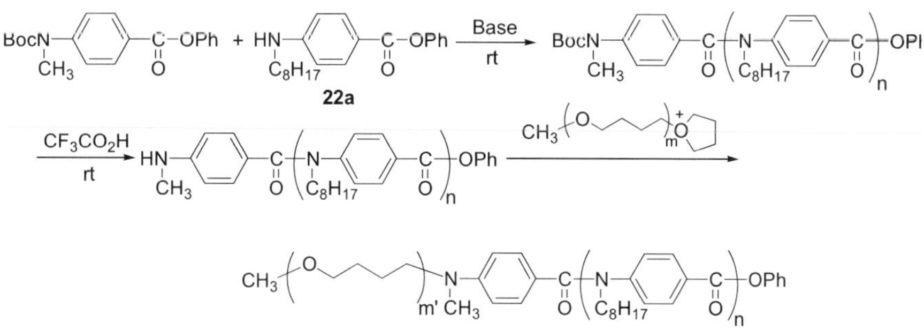

Scheme 89

Another approach to block copolymers of the aromatic polyamide and coil polymer is the macroinitiator method: the chain-growth condensation polymerization of **22a** from a macroinitiator derived from coil polymer. A diblock copolymer of polystyrene and the polyamide was synthesized with this approach (Scheme 90). First, polystyrene with a terminal carboxyl group was prepared by anionic living polymerization of styrene, followed by quenching with dry ice, and then the carboxyl group was converted to the phenyl ester by using phenol and a condensation agent. From this terminal, chain-growth polycondensation of **22a** was carried out to yield the desired block copolymer. When low molecular weight macroinitiators were used, block copolymers with low polydispersity were obtained in good yields. When high molecular weight macroinitiators were used, the homopolymer of the polyamide was contaminated. This is probably because the polymer effect of polystyrene decreased the efficiency of initiation from the macroinitiator to induce self-polycondensation of **22a** [310]. Tri-armed star polyamides were also synthesized [311].

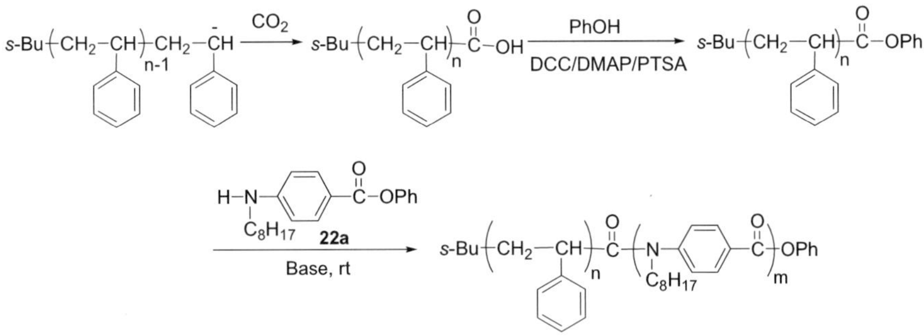

Scheme 90

The synthesis of well-defined aromatic polyester was more difficult than that of polyamide, because polyester easily undergoes transesterification. The monomer can attack the polymer ester linkage to generate the cleaved chain with the phenoxide moiety and the acyl group at both ends, leading to conventional step-growth condensation polymerization. Actually, transesterification occurred in the condensation polymerization of monomer having an active amide moiety as a good leaving group, even with a weak base such as tertiary amine at room temperature [312]. However, when the polymerization of this monomer was carried out at –30 °C with Et_3SiH, CsF, and 18-crown-6 as a base system, transesterification was almost suppressed and the molecular weight was controlled up to 7300 with low polydispersity ($M_w/M_n \leq 1.3$) (Scheme 91) [313].

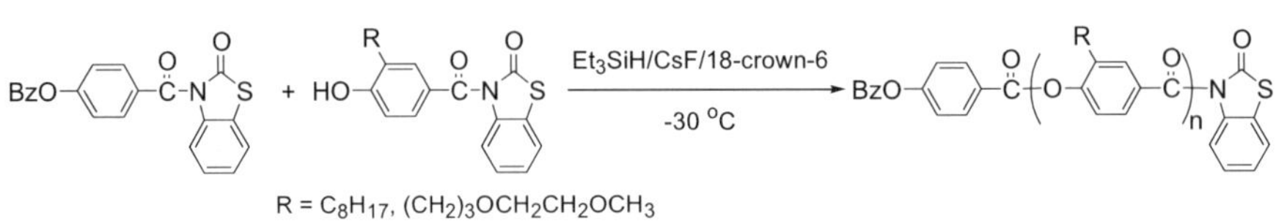

Scheme 91

From the perspective of the use of different substituent effects between the monomer and the polymer for chain-growth condensation polymerization, the synthesis of well-defined polyethers seemed to be difficult because the hydroxyl group in a monomer and the ether linkage of a polymer have similar electron-donating ability. However, monomer **24a** bearing a phenoxide moiety underwent chain-growth condensation polymerization in sulfolane at 150 °C to yield an aromatic polyether with low polydispersity ($M_w/M_n \leq 1.1$). The molecular weight was controlled up to 3500, because the polyether with molecular weight higher than that precipitated during polymerization [314]. Monomer **24b** substituted with an octyl group instead of the propyl group in **24a** also afforded an insoluble polymer in sulfolane when attempts were made to prepare higher molecular weight polymer. In other solvents, such as DMI and tetraglyme, the polymerization proceeded homogeneously, but both chain-growth and step-growth polymerization took place to give a polyether with broad molecular weight distribution [315]. The key to successful chain-growth condensation polymerization of **24** is the use of phenoxide in the monomer instead of phenol. The phenoxide moiety works as a stronger electron-donating group than the phenol moiety and the ether linkage, and the carbon attached to the fluorine in monomer **24** is strongly deactivated to prevent **24** from reacting with each other. Accordingly, **24** reacts selectively with the initiator and the polymer end group, resulting in chain-growth condensation polymerization (Scheme 92).

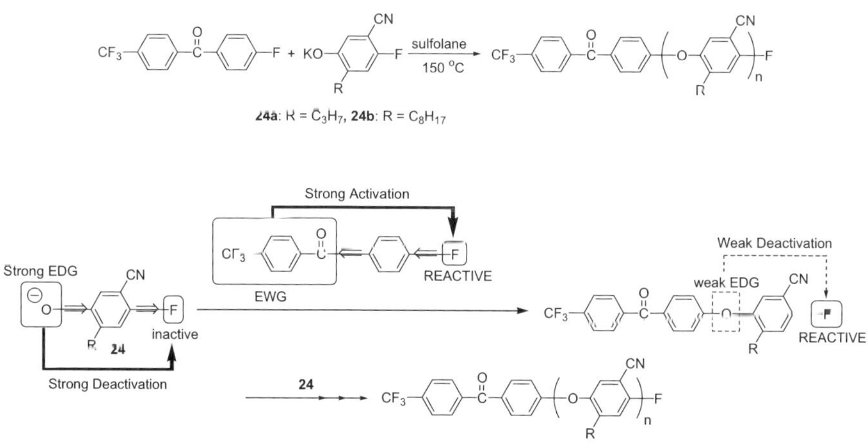

Scheme 92

An interesting thing is that the polyether with low polydispersity from chain-growth condensation polymerization possessed higher crystallinity than the one with broad molecular weight distribution from conventional step-growth condensation polymerization. The XRD pattern of the former polymer showed a stronger intensity, and the DSC profile showed the exothermic peak at 172 °C (cold crystallization) by heating from the glassy state [316]. This implies that the crystallinity of condensation polymers could be controlled by polydispersity.

Well-defined diblock copolymers of polystyrene and aromatic polyether have been recently synthesized by the combination of ATRP and this chain-growth condensation polymerization from an orthogonal initiator, which has two different initiating sites for two kinds of polymerization, with one of the initiating sites being inert with respect to polymerization from the other initiating site [317]. A polystyrene macroinitiator was first synthesized by the ATRP of styrene in the presence of an orthogonal initiator with the benzylic bromide moiety, then the terminal bromine of the polystyrene was dehalogenated with Bu_3SnH. The chain-growth condensation polymerization of **24a** was then carried out with the polystyrene macroinitiator in sulfolane at 150 °C to afford well-defined polystyrene-*b*-aromatic polyether (Scheme 93). This diblock copolymer self-assembled in THF to form spherical aggregates.

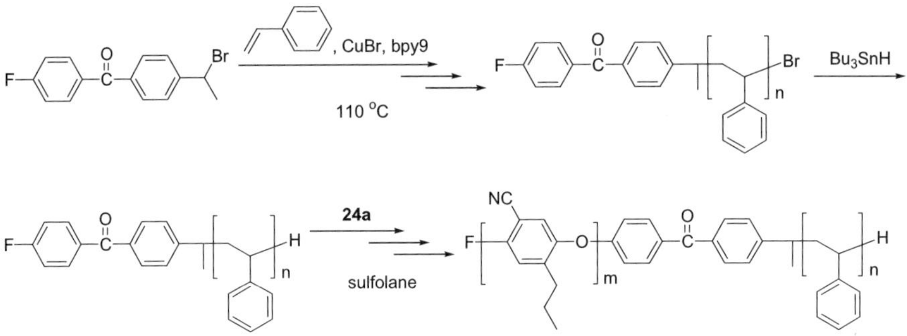

Scheme 93

The chain-growth condensation polymerization leading to aromatic polyether can be applied to the synthesis of a well-defined poly(ether sulfone) by the condensation polymerization of **25**, which is different from other monomers for chain-growth condensation polymerization in that the nucleophilic site and electrophilic site are on each benzene ring connected with an electron-withdrawing group, a sulfonyl group (Scheme 94). In the polymerization of **25** in the presence of an initiator and 18-crown-6 in sulfolane at

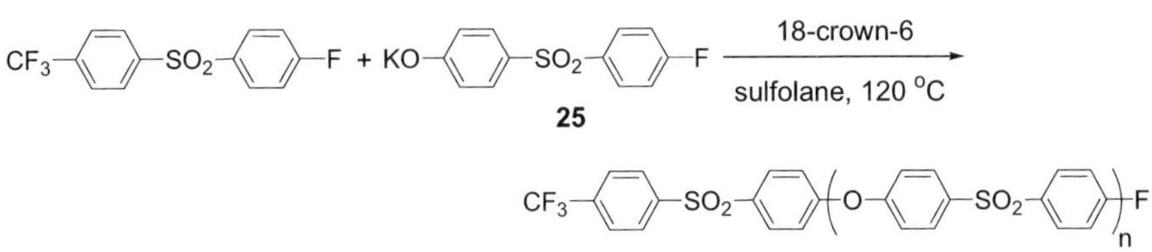

Scheme 94

120 °C, the molecular weight was controlled up to 5700 and the molecular weight distribution was less than 1.5. When the polymerization was carried out at higher feed ratio of the monomer to initiator, both chain-growth and step-growth condensation polymerization occurred [318]. This undesirable step-growth condensation polymerization is caused by transetherification of the backbone ether linkage with the monomer and/or fluoride, which is common in the condensation polymerization for poly(ether sulfone) at high temperature [225, 226]. Similar chain-growth condensation polymerization in the case of poly(ether ketone) was also reported [319].

Recently, Ueda, Rusanov, and coworkers attempted chain-growth condensation polymerization of a 4-iodophenylethyne derivative with a Pd and Cu catalyst in the presence of 1-ethynyl-2,5-dihexyl-4-(2-phenylethynyl)benzene as an initiator and tertiary amine as a base [320] (Scheme 95). The acidity of the proton of the ethynyl group in the polymer end group was expected to be higher than that of the monomer, owing to longer conjugation in the polymer chain, and to result in selective abstraction of the proton of the polymer end group with the base. However, oxidative coupling of the ethynyl group occurred as a side reaction, and poly(phenylethynylene) with low polydispersity was obtained when the feed ratio of monomer to initiator was only 1 or 2.

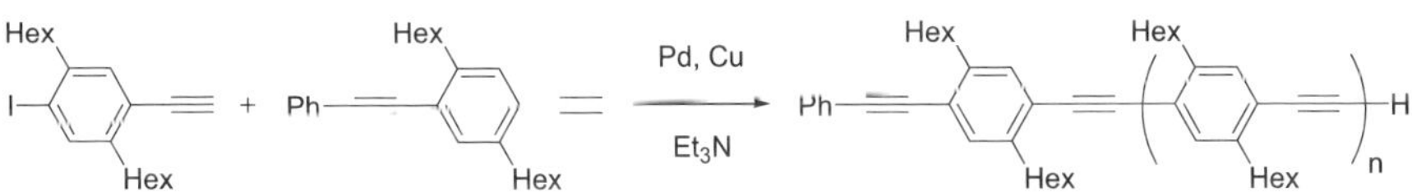

Scheme 95

7.2.2
Inductive Effect (Polymerization of *m*-Substituted Monomers)

In condensation polymerization of *meta*-substituted monomers, the inductive effect (+*I* effect) of the nucleophilic site on the reactivity of the electrophilic site at the *meta* position of the monomer is as applicable to chain-growth condensation polymerization as is the *R* effect of the *para*-substituted monomers mentioned above. Thus, ethyl 3-(alkylamino)benzoate (**26**) polymerized in the presence of LiHMDS as a base and phenyl 4-methylbenzoate as an initiator in THF at 0 °C to yield *N*-alkylated poly(*m*-benzamide)s with well-defined molecular weights and low polydispersities ($M_w/M_n \leq 1.1$) (Scheme 96). When the *N*-alkyl group is an octyl group, the M_n of the polymer was controlled up to 12 000 by the feed ratio of the monomer to initiator, and the polydispersity was kept narrow. In this polymerization, the aminyl anion of deprotonated **26** would deactivate the acyl group at the *meta* position through the strong +*I* effect, resulting in suppression of the self-polycondensation of **26**. The anion of **26** would then selectively react with

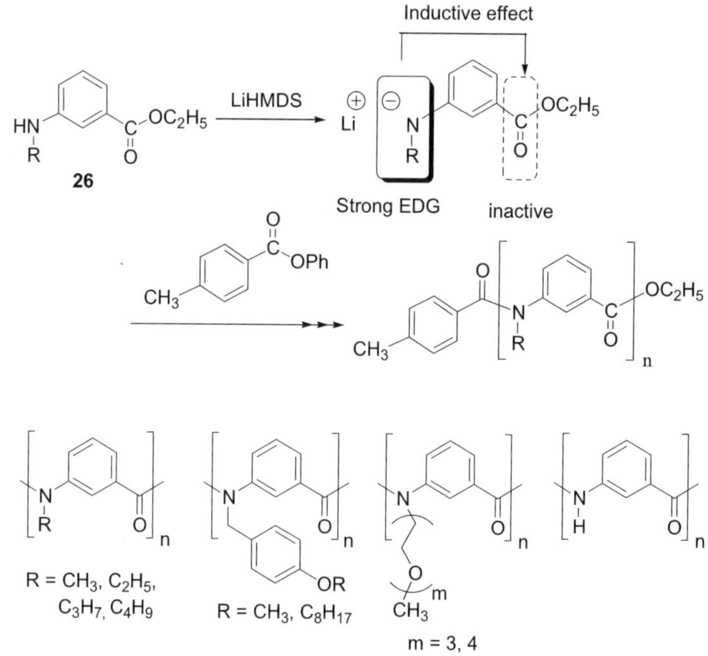

Scheme 96

an initiator and the polymer chain end, the acyl group of which is more reactive than that of the monomer with the aminyl anion, and growth would continue in a chain-polymerization manner. To support this mechanism, DFT calculations were performed. The activation energies for the propagation and self-condensation are 21.6 and 27.0 kcal mol^{-1}, respectively. On the basis of the geometries, energies, and vibrational frequencies obtained, the theoretical rate constants were then evaluated at 298.15 K and 1 atm. The reaction rate constant ($1.1 \times 10^{-3}\ s^{-1}$) for the propagation is 8.6×10^3-fold greater than that for the self-condensation ($1.3 \times 10^{-7}\ s^{-1}$), and hence is consistent with the experimental finding that propagation was observed exclusively over self-condensation; that is, chain-growth condensation polymerization of *meta*-substituted aminobenzoic ester monomers proceeded [321].

A variety of well-defined poly(*m*-benzamide)s were synthesized from the corresponding monomers (Scheme 96). All these polymers have higher solubility than the *para*-substituted counterparts [322]. Especially, the polyamides having an oligo(ethylene glycol) are soluble in water and show reversible cloud points by heating [323].

A well-defined diblock copolymer of *meta*- and *para*-substituted poly(benzamide) was also synthesized by this polymerization method (Scheme 97). Thus, ethyl 3-(octylamino)benzoate (**26a**) was polymerized with 2.2 equivalents of LiHMDS at 0 °C to give a prepolymer. A fresh feed of methyl 4-(octyl-

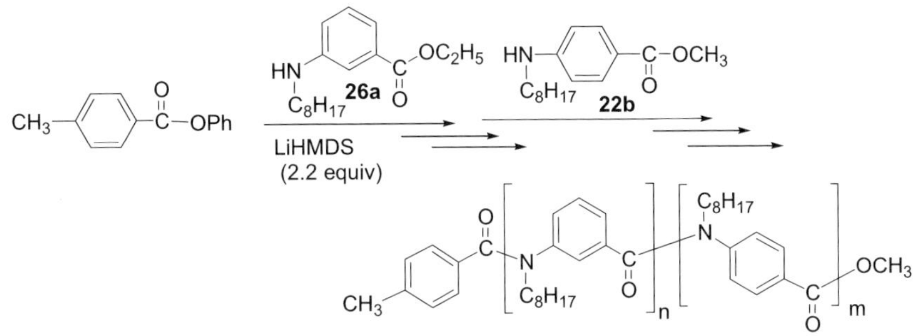

Scheme 97

amino)benzoate (**22b**) was added to the prepolymer in the reaction mixture at the same temperature to yield the block copolymer. It should be noted that excess LiHMDS in the polymerization of **26a** as the first step did not react at all with the terminal ester moiety of poly**26a**, which was able to initiate the polymerization of **22b** as the second step [321].

7.3 Transfer of Catalyst

Condensation polymerization with a catalyst can involve another mechanism for chain-growth condensation polymerization, that is, a catalyst-transfer mechanism, in which the catalyst activates the polymer end group, followed by reaction with the monomer and transfer of the catalyst to the elongated polymer end group, in a similar manner to biological condensation polymerization. We and McCullough et al. have established this mechanism for the Ni-catalyzed condensation polymerization leading to polythiophene. The polymerization of **27a** with a Ni catalyst was well known as a region-regulated synthetic method for poly(alkylthiophene)s developed by McCullough et al., but the polymers obtained possessed a broad molecular weight distribution [154]. However, when the polymerization was carried out at room temperature, with care to use the exact amount of isopropyl magnesium chloride for generation of monomer **27a** from the corresponding dehalogenated monomer, the M_n values of polymers increased in proportion to monomer conversion, with narrow polydispersities bring retained, and were controlled by the amount of the Ni catalyst; the M_n values were proportional to the feed ratio of [**27a**]$_0$/[Ni catalyst]$_0$ [324, 325]. A similar zinc monomer showed the same polymerization behavior [326]. Furthermore, the M_w/M_n ratios were around 1.1 up to M_n of 28 700 when the polymerization of **27a** was quenched with hydrochloric acid [327].

After a detailed study of the polymerization of **27a**, four important points were clarified: (1) the polymer end groups are uniform among molecules—one end group is Br and the other is H; (2) the propagating end group is a polymer–Ni–Br complex; (3) one Ni molecule forms one polymer chain; and (4) the chain initiator is a dimer of **27a** formed in situ. On the basis of these results, we have proposed a catalyst-transfer condensation polymerization mechanism (Scheme 98). Thus, Ni(dppp)Cl_2 reacts with 2 equivalents of **27a**, and the coupling reaction occurs with concomitant generation of a zero-valent Ni complex. The Ni(0) complex is not stable, and is inserted into the intramolecular C–Br bond. Another **27a** molecule reacts with this Ni, followed by the coupling reaction and transfer of the Ni catalyst to the next C–Br bond. Growth would continue in such a way that the Ni catalyst moves to the polymer end group [328].

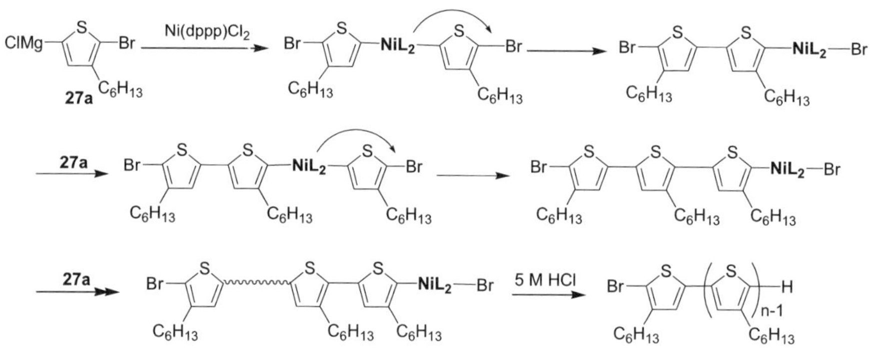

Scheme 98

In the chain-growth condensation polymerization mechanism based on the substituent effect, it would sometimes be difficult to control the molecular weight in the region of very high molecular weights because the reaction between monomers, which leads to self-polycondensation, cannot be completely suppressed at very low concentration of the propagating group. On the other hand, this catalyst-transfer condensation polymerization is different in that the monomers essentially do not react with each other but react with only the polymer end group if the catalyst moves to the polymer end group. This means that the molecular weight of a polymer ideally can be controlled no matter how large it is.

On the basis of the nature of the polythiophene end group containing the Ni complex, functional groups can be introduced to one or both ends of the polymer by Grignard reagents. Allyl, ethynyl, and vinyl Grignard reagents afford monofunctionalized polythiophenes, whereas aryl and alkyl Grignard reagents yield difunctionalized polythiophenes. By utilizing the proper protecting groups, hydroxyl, formyl, and amino groups can also be incorporated into the polymer ends [329].

Regioregular polythiophenes with various side chains containing ether, ester, and thiol units had been synthesized, and showed interesting properties [330]. However, the polydispersity of these polymers was not controlled. We studied chain-growth polymerization of a thiophene monomer **27b** with an oxyethylene side chain. The polymerization of **27b** depended on the ligands of the Ni catalyst, and poly**27b** with the lowest polydispersity was obtained when dppe was used as the ligand (Scheme 99) [331]. Catala and coworkers reported that the polymerization of a thiophene monomer **27c** containing an ester moiety on the side chain with Ni(dppp)Cl_2 also proceeded in a chain-growth polymerization manner to yield head-to-tail poly**27c** with a narrow polydispersity, and that the molecular weight of the polymer was controlled by the feed ratio of the monomer to Ni(dppp)Cl_2 (Scheme 100) [332]. The block copolymer of poly(3-hexylthiophene) and poly(3-dodecylthiophene) [325] and the block copolymer of poly(3-hexylthiophene) and poly**27b** [333] were synthesized by the successive chain-growth polymerization of different thiophene monomers.

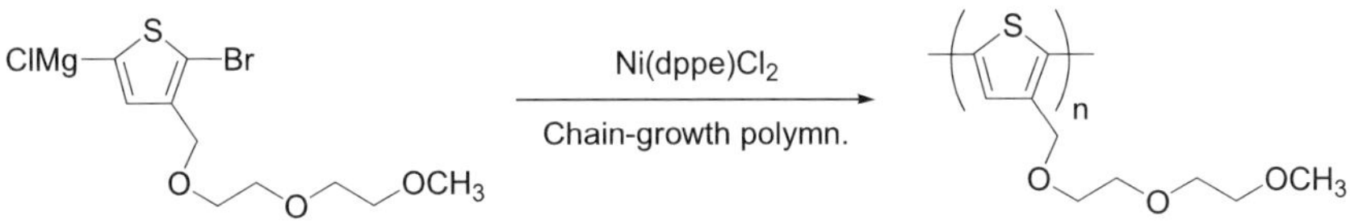

27b

Scheme 99

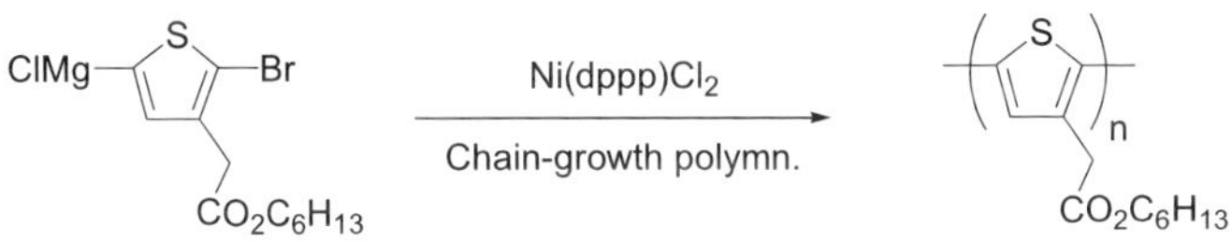

27c

Scheme 100

McCullough and coworkers had synthesized block copolymers of poly(3-hexylthiophene) and polystyrene or PMA by ATRP of the vinyl monomer from a polythiophene macroinitiator, which was prepared in several steps [334]. After the development of catalyst-transfer condensation polymerization of polythiophene, the block copolymer of polythiophene and PMA could be prepared more easily. As mentioned above, the vinyl-terminated polythiophene was first prepared. The vinyl group was converted to the 2-hydroxyethyl group by hydroboration, followed by esterification with 2-bromopropionyl bromide to give a macroinitiator for ATRP (Scheme 101) [335]. Recently, Dai, Su, and coworkers synthesized the block copolymer of polythiophene and polypyridine from the vinyl-terminated polythiophene (Scheme 102) [336]. Meijer and coworkers used an allyl-terminated polythiophene to synthesize a block copolymer of poly(3-hexylthiophene) and polyethylene: the ring-opening metathesis polymerization of cyclooctene in the presence of the allyl-terminated polythiophene was followed by hydrogenation (Scheme 103) [337].

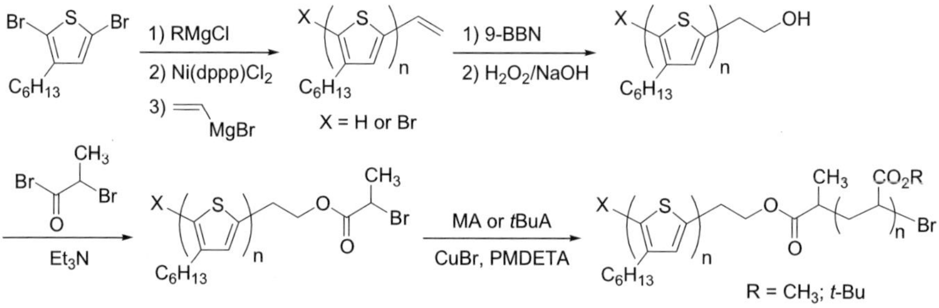

Scheme 101

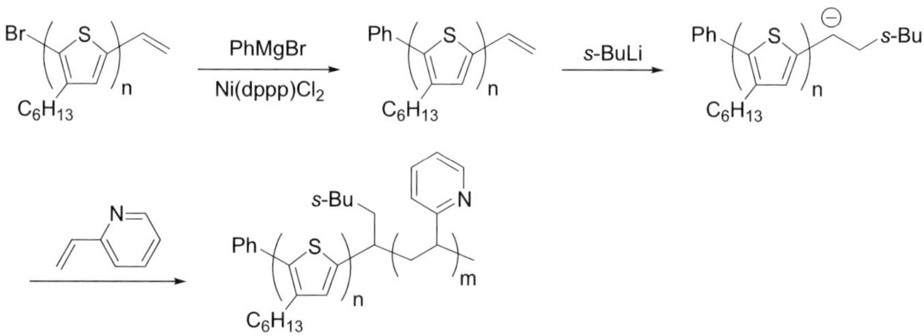

Scheme 102

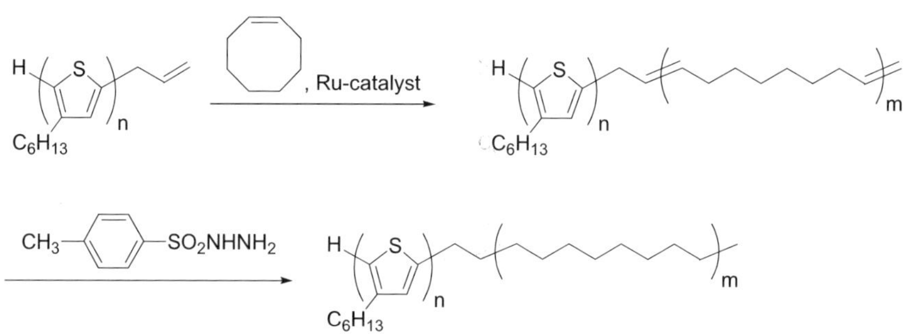

Scheme 103

It is important to clarify whether catalyst-transfer condensation polymerization is specific to polythiophene, or whether it is generally applicable to the synthesis of well-defined π-conjugated polymers. We investigated the synthesis of poly(*p*-phenylene), to see whether a monomer **28** containing no heteroatom in the aromatic ring would undergo catalyst-transfer polymerization. However, all polymers obtained in the polymerization with Ni(dppp)Cl_2, Ni(dppe)Cl_2, or Ni(dppf)Cl_2 possessed low molecular weights and broad polydispersities. Nevertheless, we found that LiCl was necessary for optimizing the chain-growth condensation polymerization leading to poly(*p*-phenylene) with low polydispersity, and that the molecular weight was controlled by the feed ratio of **28** to the Ni catalyst (Scheme 104) [338].

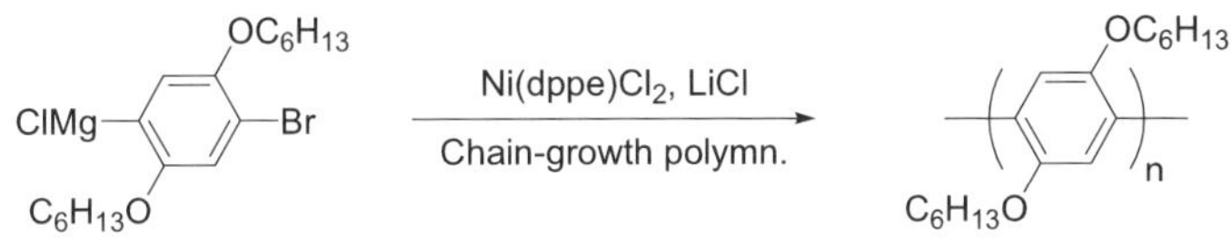

Scheme 104

Furthermore, we investigated chain-growth Suzuki–Miyaura coupling polymerization. In this polymerization, stable arylpalladium(II) halide complex could be used as an externally added initiator, and the aryl group of the complex would serve as an initiator unit of the polymer. The polymerization of a fluorene monomer **29** was carried out in the presence of $^{t}Bu_3PPd(Ph)Br$ as a catalyst to yield polyfluorene with a narrow polydispersity (Scheme 105). The molecular weight of the obtained polymer increased linearly in proportion to the conversion of monomer with low polydispersity throughout the polymerization, and also increased linearly in proportion to the feed ratio of **29** to the initiator up to 17 700 with low polydispersity, indicating that this Suzuki–Miyaura coupling condensation polymerization proceeded through a chain-growth polymerization mechanism [339], as shown in the model reactions [340–342]. The MALDI-TOF mass spectrum of the obtained polyfluorene showed that all the polymers bore the phenyl group at one end. This observation strongly supported the view that $^{t}Bu_3PPd(Ph)Br$ served as an initiator. The catalyst-transfer polymerization with an externally added initiator would provide surface-grafted π-conjugated polymers with defined molecular weight and low polydispersity by the use of a surface-bound Pd complex initiator.

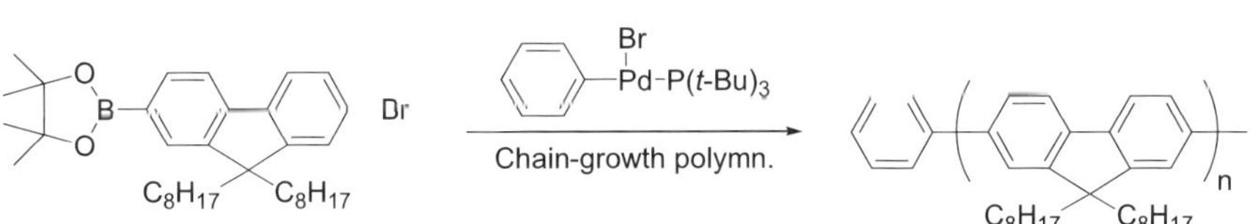

Scheme 105

7.4 Biphasic System

A quite different approach to chain-growth condensation polymerization is phase-transfer-catalyzed polycondensation of solid monomer dispersed in an organic solvent. We expected that the solid phase of the monomer would prevent the reaction of monomers with each other, and that the monomer dissolved in an organic solvent with the aid of a PTC in a certain amount would react with an initiator and the polymer end group in the solution phase. The solid monomer, potassium 4-bromomethyl-2-octyloxybenzoate, was polymerized in the presence of 18-crown-6 as a PTC and 4-nitrobenzyl bromide as the initiator in acetone (Scheme 106) [343]. The polymerization successfully proceeded by chain growth to yield polyesters with M_w/M_n less than 1.3 until the feed of the monomer to the initiator reached 15. With a ratio of 20 or above, the control of polymerization was not perfect: polymers having M_n values close to the calculated ones were obtained as well as oligomers without the initiator unit. Similar chain-growth polymerization was also attained with tetrabutylammonium iodide as the PTC instead of 18-crown-6 [344].

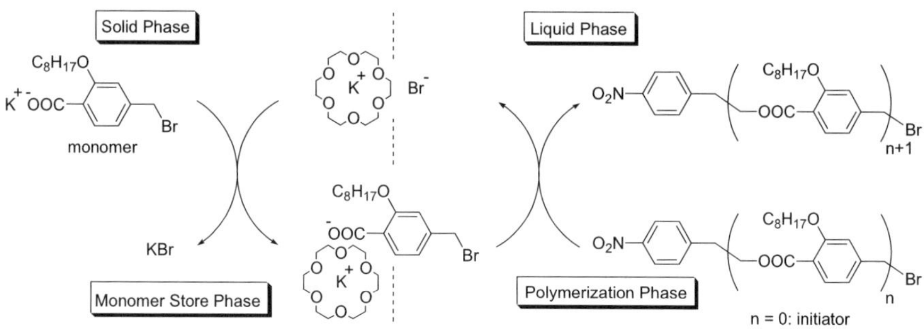

Scheme 106

8 Conclusion and Future Outlook

We have described reaction control in condensation polymerization during a decade, examples of which have been far more than we thought at the first outline. If reaction control of *addition polymerization*, for example, were reviewed, so many sections would not be described, although a huge number of works have been carried out in a certain section: for example, living radical polymerization. A reason why many kinds of control have been investigated in condensation polymerization seems to be that condensation polymerization can involve considerably more kinds of reactions than addition polymerization. Most of the substitution reactions developed in organic chemistry can apply to condensation polymerization. Therefore, further new condensation polymerizations will be developed along with the development of organic chemistry.

References

1. Carothers WH (1936) Trans Faraday Soc 32:39
2. Flory PJ (1936) J Am Chem Soc 58:1877
3. Flory PJ (1946) Chem Rev 39:137
4. Yamazaki N, Higashi F (1974) J Polym Sci B Polym Lett 12:185
5. Yamazaki N, Higashi F (1974) J Polym Sci A Polym Chem 12:2149
6. Ueda M, Kameyama A, Hashimoto K (1988) Macromolecules 21:19
7. Whinfield JR, Dickson JT (1949) US Patent 2465319 19490322
8. Pilati F (1989) Polyesters. In: Allen G (ed) Comprehensive polymer science, vol 5. Pergamon, Oxford, p 275
9. Banach TE, Buti C, Colonna M, Fiorini M, Marianucci E, Messori M, Pilati F, Toselli M (2001) Polymer 42:7511
10. Banach TE, Colonna M (2001) Polymer 42:7514
11. Colonma M, Banach TE, Berti C, Fiorino M, Marianucci E, Messori M, Pilati F, Toselli M (2003) Polymer 44:4773
12. Kricheldorf HR, Masri MA, Lomadze N, Schwartz G (2005) Macromolecules 38:9085
13. Thiele UK (2001) Int J Polym Mater 5:387
14. Zimmermann H (1962) Faserforsch Textiltech 13:481
15. Yoda K (1971) Kogyo Kagaku Zasshi 74:1476
16. Rafler G, Reinisch G, Bonatz E (1974) Acta Chim 81:253 (Budapest)
17. Tomita K, Ida H (1975) Polymer 16:185
18. Laserova R, Dimov K (1976) Angew Makromol Chem 55:1
19. Rafler G, Tesch F, Kunath D (1988) Acta Polym 39:315
20. Otera J, Dan-oh N, Nozaki H (1991) J Org Chem 56:5307
21. Otera J, Kawada K, Yano T (1996) Chem Lett, p 225
22. Takahashi H, Hayakawa T, Ueda M (2000) Chem Lett, p 684
23. Velmathi S, Nagahata R, Sugiyama J, Takeuchi K (2005) Macromol Rapid Commun 26:1163
24. Ishihara K, Ohhara S, Yamamoto H (2000) Science 290:1140
25. Ishihara K, Ohara S, Yamamoto H (1996) J Org Chem 61:4196
26. Ishihara K, Ohara S, Yamamoto H (2000) Macromolecules 33:3511
27. Takasu A, Iio Y, Oishi Y, Narukawa Y, Hirabayashi T (2005) Macromolecules 38:1048
28. Takasu A, Iio Y, Miura T, Hirabayashi T (2005) Polym J 37:946
29. Tsonis CP (1989) In: Allen G, Bevington JC (eds) Comprehensive polymer science, vol 5, chap 26. Pergamon, Elmsford, p 455
30. Tsonis CP (1996) In: Salamone JC (ed) Polymeric materials encyclopedia. CRC, Boca Raton, p 407
31. Olah GA (1993) Angew Chem Int Ed Engl 32:767
32. Gassman PG, Tidwell TT (1983) Acc Chem Res 16:279
33. Tidwell TT (1984) Angew Chem Int Ed Engl 23:20
34. Creary X (1991) Chem Rev 91:1625
35. Klumpp DA (2001) Recent Res Dev Org Chem 5:194
36. Yamazaki T, Saito S, Ohwada T, Sudo K (1995) Tetrahedron Lett 36:5749
37. Klumpp DA, Yeung KY, Prakash GKS, Olah GA (1998) Synlett, p 918
38. Klumpp DA, Yeung KY, Prakash GKS, Olah GA (1998) J Org Chem 63:4481
39. Nenajdenko VG, Shevchenko NE, Balenkova ES, Alabugin IV (2003) Chem Rev 103:229
40. Colquhoun HM, Zolotukhin MG, Khalilov LM, Dzhemilev UM (2001) Macromolecules 34:1122

41. Zolotukhin MG, Hernández MCG, Lopez AM, Fomina L, Cedillo G, Nogales A, Equerra T, Rueda D, Colquhoun HM, Fromm KM, Ruiz-Treviño A, Ree M (2006) Macromolecules 39:4696
42. Smet M, Schacht E, Dehaen W (2002) Angew Chem Int Ed 41:4547
43. Zolotukhin MG, Fomina L, Salcedo R, Sansores LE, Colquhoun HM, Khalilov LM (2004) Macromolecules 37:5140
44. Peña ER, Zolotukhin MG, Fomine S (2004) Macromolecules 37:6227
45. Zolotukhin MG, Fomine S, Lazo LM, Salcedo R, Sansores LE, Cedillo GG (2005) Macromolecules 38:6005
46. Saito S, Saito S, Ohwada T, Shudo K (1991) Chem Pharm Bull 39:2718
47. Fu Y, Vandendriessche A, Dehaen W, Smet M (2006) Macromolecules 39:5183
48. Zolotukhin M, Fomine S, Lazo LM, Salcedo R, Khalilov L (2004) Chem Commun, p 1030
49. Peña ER, Zolotukhin M, Fomine S (2005) Polymer 46:7494
50. Lira AL, Zolotukhin MG, Fomina L, Fomine S (2007) Macromol Theory Simul 16:227
51. Diaz AM, Zolotukhin MG, Fomine S, Salcedo R, Manero O, Cedillo G, Velasco VM, Guzman MT, Fritsch D, Khalizov AF (2007) Macromol Rapid Commun 28:183
52. Yamamoto T, Yamamoto A (1997) Chem Lett, p 353
53. Yamamoto T, Hayashi Y, Yamamoto A (1978) Bull Chem Soc Jpn 51:2091
54. Yamamoto T (1992) Prog Polym Sci 17:1153
55. Schlüter AD, Wegner G (1993) Acta Polym 44:53
56. Drent E, Budzelaar HM (1996) Chem Rev 96:663
57. Schuster M, Blechert S (1997) Angew Chem Int Ed Engl 36:2036
58. Wolfe JP, Wagaw S, Marcoux JF, Buchwald SL (1988) Acc Chem Res 31:805
59. Hartwig JF (1998) Angew Chem Int Ed 37:2046
60. Yang BY, Buchwald SL (1999) Organomet Chem 576:125
61. Kanbara T, Honma A, Hasegawa K (1996) Chem Lett, p 1135
62. Hartwig JF, Richards S, Barañano D, Paul F (1996) J Am Chem Soc 118:3626
63. Kanbara T, Izumi K, Narise T, Hasegawa K (1998) Polym J 30:66
64. Goodson FE, Hartwig JF (1998) Macromolecules 31:1700
65. Kanbara T, Izumi K, Nakadani Y, Narise T, Hasegawa K (1997) Chem Lett, p 1185
66. Kanbara T, Nakadani Y, Hasegawa K (1999) Polym J 31:206
67. Kanbara T, Miyazaki Y, Hasegawa K, Yamamoto T (2000) J Polym Sci A Polym Chem 38:4194
68. Kanbara T, Imayasu T, Hasegawa K (1998) Chem Lett, p 1185
69. Al-Hussaini AS, Klapper M, Pakula T, Müllen K (2004) Macromolecules 37:8269
70. Ward RE, Meyer TY (2003) Macromolecules 36:4368
71. Kanbara T, Ohshima M, Imayasu T, Hasegawa K (1988) Macromolecules 31:8725
72. Jung BJ, Lee JI, Chu HY, Do LM, Shim HK (2002) Macromolecules 35:2282
73. Zhang XX, Sadighi JP, Mackewitz TW, Buchwald SL (2000) J Am Chem Soc 122:7606
74. Aranyos A, David WO, Kiyomori A, Wolfe JP, Sadighi JP, Buchwald SL (1999) J Am Chem Soc 121:4369
75. Shibasaki Y, Ueda M (2002) Chem Lett, p 794
76. Goyal M, Nagahata R, Sugiyama J, Asai M, Ueda M, Takeuchi K (1999) Polymer 40:3237
77. Goyal M, Nagahata R, Sugiyama J, Asai M, Ueda M, Takeuchi K (2009) Polymer 41:2289
78. Ishii H, Goyal M, Ueda M, Takeuchi K, Asai M (2001) Macromol Rapid Commun 22:376
79. Okuyama K, Sugiyama J, Nagahata R, Asai M, Ueda M, Takeuchi K (2003) J Mol Catal A Chem 203:21

80. Okuyama K, Sugiyama J, Nagahata R, Asai M, Ueda M, Takeuchi K (2003) Macromolecules 36:6953
81. Okuyama K, Sugiyama J, Nagahata R, Asai M, Ueda M, Takeuchi K (2003) Green Chem 5:563
82. Krieble RH, Burkhard RH (1947) J Am Chem Soc 69:2689
83. Hopf H, Egli R (1969) Kunstst Plast 6:433
84. Li Y, Kawakami Y (1999) Macromolecules 32:3540
85. Li Y, Kawakami Y (1999) Macromolecules 32:6871
86. Li Y, Kawakami Y (1999) Macromolecules 32:8768
87. Zhang R, Mark JE, Pinhas AR (2000) Macromolecules 33:3508
88. Rubinsztajn S, Cella JA (2005) Macromolecules 38:1061
89. Ueda M (1999) Prog Polym Sci 24:699
90. Ueda M, Kameyama A, Hashimoto K (1988) Macromolecules 21:19
91. Ueda M, Morosumi T, Kakuta M, Sato R (1990) Polym J 22:733
92. Wang C, Nakamura S (1994) J Polym Sci A Polym Chem 32:413
93. Wang C, Nakamura S (1994) J Polym Sci A Polym Chem 32:1255
94. Wang C, Nakamura S (1995) J Polym Sci A Polym Chem 33:2027
95. Wang C, Nakamura S (1995) J Polym Sci A Polym Chem 33:2157
96. Wang C, Nakamura S (1996) J Polym Sci A Polym Chem 34:493
97. Wang C, Takayama T, Nakamura S (1996) J Polym Sci A Polym Chem 34:755
98. Hayakawa T, Morishita T, Okazaki M, Ueda M, Takeuchi K, Asai M (2000) J Polym Sci A Polym Chem 38:3875
99. Okazaki M, Hayakawa T, Ueda M, Takeuchi K, Asai M (2001) J Polym Sci A Polym Chem 39:78
100. Takasu A, Shibata Y, Narukawa Y, Hirabayashi T (2007) Macromolecules 40:151
101. Yao D, Li G (2006) Chin Chem Lett, p 1611
102. Yao D, Li G, Zong M (2007) Chem Lett 36:574
103. Kobayashi S, Kashiwa K, Kawasaki T, Shoda S (1991) J Am Chem Soc 113:3079
104. Kobayashi S, Wen X, Shoda S (1996) Macromolecules 29:2698
105. Kobayashi S, Shimada J, Kashiwa K, Shoda S (1992) Macromolecules 25:3237
106. Fuijita M, Shoda S, Kobayashi S (1998) J Am Chem Soc 120:6411
107. Okamoto E, Kiyosada T, Shoda S, Kobayashi S (1997) Cellulose 4:161
108. Kobayashi S, Ohmae M (2006) Adv Polym Sci 194:159
109. Kline BJ, Beckman EJ, Russell AJ (1998) J Am Chem Soc 120:9475
110. Uyama H, Kobayashi S (1994) Chem Lett, p 1687
111. Kobayashi S, Uyama H, Kimura S (2001) Chem Rev 101:3793
112. Tsujimoto T, Uyama H, Kobayashi S (2001) Biomacromolecules 2:29
113. Uyama H, Inada K, Kobayashi S (2001) Macromol Biosci 1:40
114. Uyama H, Inada K, Kobayashi S (1999) Macromol Rapid Commun 20:171
115. Chaudhary AK, Lopez J, Becjmann EJ, Russell AJ (1997) Biotechnol Prog 13:318
116. Kim DY, Dordick JS (2001) Biotechnol Bioeng 76:200
117. Park O, Kim DY, Dordick JS (2000) Biotechnol Bioeng 70:208
118. Uyama H, Klegraf E, Wada S, Kobayashi S (2000) Chem Lett, p 800
119. Morimoto T, Murakami N, Nagatsu A, Sakakibara J (1994) Chem Pharm Bull 42:751
120. Patil DR, Rethwisch DG, Dordick JS (1991) Biotechnol Bioeng 37:639
121. Kumar A, Kulshrestha AS, Gao W, Gross RA (2003) Macromolecules 36:8219
122. Nomura N, Tsurugi K, Okada M (1999) J Am Chem Soc 121:7269
123. Nomura N, Yoshida N, Tsurugi K, Aoi K (2003) Macromolecules 36:3007
124. Komiyama S, Nomura N, Saba M, Kondo T (2006) Polym Prepr Jpn 55:227
125. Hay AS, Blanchard HS, Enders GF, Eustance JW (1959) J Am Chem Soc 81:6335

126. Hay AS (1998) J Polym Sci A Polym Chem 36:505
127. Kobayashi S, Shoda S, Uyama H (1995) Adv Polym Sci 121:1
128. Kobayashi S (1999) J Polym Sci A Polym Chem 37:3041
129. Gross RA, Kumar A, Kalra B (2001) Chem Rev 101:2097
130. Mita N, Tawaki S, Uyama H, Kobayashi S (2002) Chem Lett, p 402
131. Mita N, Tawaki S, Uyama H, Kobayashi S (2004) Bull Chem Soc Jpn 77:1523
132. Kim YJ, Uyama H, Kobayashi S (2003) Macromolecules 36:5058
133. Higashimura H, Fujisawa K, Moro-oka Y, Kubota M, Shiga A, Terahara A, Uyama H, Kobayashi S (1998) J Am Chem Soc 120:8529
134. Higashimura H, Kubota M, Shiga A, Fujisawa K, Moro-oka Y, Uyama H, Kobayashi S (2000) Macromolecules 33:1986
135. Kitajima N, Fujisawa K, Fujimoto C, Moro-oka Y, Hashimoto S, Kitagawa T, Toriumi K, Tatsumi K, Nakamura A (1992) J Am Chem Soc 114:1277
136. Mahapatra S, Halfen JA, Wilkinson EC, Que L Jr, Tolman WB (1994) J Am Chem Soc 116:9785
137. Kitajima N, Fujisawa K, Moro-oka Y, Toriumi K (1989) J Am Chem Soc 111:8975
138. van Dort HM, Hoefs CAM, Magré EP, Schöpf AJ, Yntema K (1968) Eur Polym J 4:275
139. Higashimura H, Fujisawa K, Moro-oka Y, Namekawa S, Kubota M, Shiga A, Uyama H, Kobayashi S (2000) Macromol Rapid Commun 21:1121
140. Cheng SZD, Wunderlich B (1987) Macromolecules 20:1630
141. Shibasaki Y, Nakamura M, Ishimaru R, Kondo JN, Domen K, Ueda M (2004) Macromolecules 28:9657
142. Shibasaki Y, Nakamura M, Kondo JN, Ueda M (2006) Macromol Symp 245–246:87
143. Ueda M, Abe T, Awano H (1992) Macromolecules 25:5125
144. Yamamoto K, Tsuchida E, Nishiide H, Jikei M, Oyaizu K (1993) Macromolecules 26:3432
145. Okada T, Ogata T, Ueda M (1996) Macromolecules 29:7645
146. Noji M, Nakajima M, Koga K (1994) Tetrahedron Lett 35:7983
147. Amou S, Takeuchi K, Asai M, Niizeki K, Okada T, Seino M, Haba O, Ueda M (1999) J Polym Sci A Polym Chem 37:3702
148. Sasada Y, Shibasaki Y, Suzuki M, Ueda M (2003) Polymer 44:355
149. Tsuchiya K, Ishii H, Shibasaki Y, Ando S, Ueda M (2004) Macromolecules 37:4794
150. Anderson MR, Selse D, Berggren M, Järvinen H, Hjertberg T, Inganäs O, Wennerström O, Österholm JE (1994) Macromolecules 27:6503
151. Haba O, Hayakawa T, Ueda M, Kawaguchi H, Kawazoe T (1998) React Funct Polym 37:163
152. Leclerc M, Diaz FM, Wegner G (1989) Makromol Chem 190:3105
153. Amou S, Haba O, Shirato K, Hayakawa T, Ueda M, Takeuchi K, Asai M (1999) J Polym Sci A Polym Chem 37:1943
154. McCullough RD, Lowe RD, Jayaraman M, Anderson DL (1993) J Org Chem 58:904
155. Chen TA, Wu X, Rieke RD (1995) J Am Chem Soc 117:233
156. Son S, Dodabalapur A, Lovinger AJ, Galvin ME (1995) Science 269:36
157. Liao L, Ding L, Karasz FE, Pang Y (2004) J Polym Sci A Polym Chem 42:303
158. Cassano T, Tommasi R, Babudri F, Cardone A, Farinola GM, Naso F (2003) Synth Met 139:827
159. Drury A, Maier S, Ruther M, Blau WJ (2003) J Mater Chem 13:485
160. Holzer W, Penzkofer A, Gong SH, Bleyer A, Bradley CDC (1996) Adv Mater 8:974
161. Katayama H, Nagao M, Moriguchi R, Ozawa F (2003) J Organomet Chem 676:49
162. Katayama H, Nagao M, Nishimura T, Matsui Y, Umeda K, Akamatsu K, Tsuruoka T, Nawafune H, Ozawa F (2005) J Am Chem Soc 127:4350

163. Katayama H, Nagao M, Nishimura T, Matsui Y, Fukuse Y, Wakioka M, Ozawa F (2006) Macromolecules 39:2039
164. Nagao M, Asano K, Umeda K, Katayama H, Ozawa F (2005) J Org Chem 70:10511
165. Katayama H, Ozawa F, Matsumiya Y, Watanabe H (2006) Polym J 38:184
166. Habaue S, Seko T, Okamoto Y (2002) Macromolecules 35:2437
167. Habaue S, Seko T, Isonaga M, Ajiro H, Okamoto Y (2003) Polym J 35:592
168. Habaue S, Ajiro H, Yoshii Y, Hirasa T (2004) J Polym Sci A Polym Chem 42:4258
169. Habaue S, Seko T, Okamoto Y (2003) Macromolecules 36:2604
170. Habaue S, Murakami S, Higashimura H (2005) J Polym Sci A Polym Chem 43:5878
171. Temma T, Habaue S (2005) J Polym Sci A Polym Chem 43:6287
172. Temma T, Hatano B, Habaue S (2006) Polymer 47:1845
173. Oishi M, Kawakami Y (1999) Org Lett 1:549
174. Oishi M, Kawakami Y (2000) Macromoleculoes 33:1960
175. Seino M, Oishi M, Imae I, Kawakami Y (2002) Polym J 34:43
176. Oishi M, Moon JY, Janvikul W, Kawakami Y (2001) Polym Int 50:135
177. Zhou D, Kawakami Y (2005) Macromoleculoes 38:6902
178. Li Y, Seino M, Kawakami Y (2000) Macromolecules 33:5312
179. Kricheldorf HR, Eggerstedt S (1998) J Polym Sci A Polym Chem 36:1373
180. Kricheldorf HR, Langanke D (1999) Macromol Chem Phys 200:1183
181. Kricheldorf HR, Langanke D, Spickermann J, Schmidt M (1999) Macromolecules 32:3559
182. Kricheldorf HR, Schwarz G (2003) Macromol Rapid Commun 24:359
183. Kricheldorf HR (2003) Macromoleculoes 36:2302
184. Kricheldorf HR, Böhme S, Schwarz G, Schultz CL (2002) Macromol Rapid Commun 23:803
185. Kricheldorf HR, Böhme S, Schwarz G (2006) Macromolecules 39:3210
186. Kricheldorf HR, Böhme S, Schwarz G, Krüger RP, Schultz G (2001) Macromolecules 34:8886
187. Kricheldorf HR, Böhme S, Schwarz G (2001) Macromolecules 34:8879
188. Percec V, Chu P, Kawasanai M (1994) Macromolecules 27:4411
189. Chu F, Hanker CJ, Pomery PJ, Hall DJT (1997) J Polym Sci A Polym Chem 35:1627
190. Gooden JK, Gross MC, Mueller A, Stefanescu AD, Wooley KL (1998) J Am Chem Soc 120:10180
191. Dusek K, Somvarsky J, Smrekova M, Simansiek JWJ, Wilezek Z (1999) Polym Bull 42:489 (Berlin)
192. Park D, Feast WJ (2001) Macromolecules 34:2048
193. Martinez CA, Hay AS (1997) J Polym Sci A Polym Chem 35:2015
194. Burgath A, Sunder A, Frey H (2000) Macromol Chem Phys 201:782
195. Kricheldorf HR, Vakhtangishvili L, Schwarz G, Krüger RP (2003) Macromolecules 36:5554
196. Kricheldorf HR, Fritsch D, Vakhtangishvili L, Schwarz G (2003) Macromolecules 36:4337
197. Kricheldorf HR, Hobzova R, Vakhtangishvili L (2005) Macromolecules 38:1738
198. Kricheldorf HR, Vakhtangishvili L, Schwarz G, Prosene M (2005) J Polym Sci A Polym Chem 43:6233
199. Kricheldorf HR, Vakhtangishvili L, Schwarz G (2004) J Polym Sci A Polym Chem 42:5725
200. Kricheldorf HR, Hobzova R, Vakhtangishvili L, Schwarz G (2005) Macromolecules 38:4630

201. Kricheldorf HR, Hobzova R, Vakhtangishvili L, Schwarz G (2005) Macromol Chem Phys 206:2133
202. Kricheldorf HR, Lomadze M, Polekfa C, Schwarz G (2006) Macromolecules 39:2107
203. Kricheldorf HR, Schellenberg J, Schwarz G (2006) Macromol Chem Phys 207:1556
204. Kricheldorf HR, Schellenberg J, Schwarz G (2006) Macromolecules 39:6445
205. Kricheldorf HR, Schellenberg J, Schwarz G (2006) J Polym Sci A Polym Chem 44:4546
206. Kricheldorf HR, Bornhorst K, Schwarz G (2007) Macromolecules 40:199
207. Brunelle DJ, Boden EP, Shannon TG (1990) J Am Chem Soc 112:2399
208. Brunelle DJ, Shannon TG (1991) Macromolecules 24:3035
209. Nagahata R, Sugiyama J, Goyal M, Asai M, Ueda M, Takeuchi K (2000) Polym Adv Technol 11:727
210. Nagahata R, Sugiyama J, Goyal M, Asai M, Ueda M, Takeuchi K (2000) Polym J 32:854
211. Brunelle DJ, Bradt JE, Serth-Guzzo J, Takekoshi T, Evans TL, Pearce EJ, Wilson PR (1998) Macromolecules 31:4782
212. Nagahata R, Sugiyama J, Goyal M, Asai M, Ueda M, Takeuchi K (2000) J Polym Sci A Polym Chem 38:3360
213. Nagahata R, Sugiyama J, Velmathi S, Nakao Y, Goto M, Takeuchi K (2004) Polym J 36:483
214. Jiang H, Qi Y, Chen T, Bo S, Ding M, Gao L, Xu J, He T, Zhang Y (2000) Macromol Chem Phys 201:2385
215. Hall AJ, Hodge P, McGrail CS, Rickerby J (2000) Polymer 41:1239
216. Hodge P, Yang Z, Ben-Haida A, McGril CS (2000) J Mater Chem 10:1533
217. Chan KP, Wang Y, Hay AS, Hronowski XL, Cotter RJ (1995) Macromolecules 28:6705
218. Ding Y, Hay AS (1996) Macromolecules 29:3090
219. Ben-Haida A, Colquhoun HM, Hodge P, Williams DJ (2000) J Mater Chem 10:2011
220. Colquhoun HM, Sestiaa LG, Zolotukhin MG, Williams DJ (2000) Macromolecules 33:8907
221. Colquhoun HM, Sestiaa LG, Williams DJ, Zolotukhin MG (2002) Macromol Rapid Commun 23:634
222. Baxter I, Colquhoun HM, Hodge P, Kohnke FH, Lewis DF, Williams DJ (2000) J Mater Chem 10:309
223. Colquhoun HM, Zolotukhin MG, Sestiaa LG, Aricó F, Zhu Z, Hodge P, Ben-Haida A, Williams DJ (2003) J Mater Chem 13:1504
224. Wang Y, Chan KP, Hay AS (1996) J Polym Sci A Polym Chem 34:375
225. Ben-Haida A, Colquhoun HM, Hodge P, Williams DJ (2006) Macromolecules 39:6467
226. Ben-Haida A, Baxter I, Colquhoun HM, Hodge P, Kohnke FH, Williams DJ (1997) Chem Commun, p 1533
227. Ben-Haida A, Baxter I, Colquhoun HM, Hodge P, Kohnke FH, Williams DJ (1998) Chem Commun, p 2213
228. Colquhoun HM, Lewis DF, Hodge P, Ben-Haida A, Williams DJ, Baxter I (2002) Macromolecules 35:6875
229. Colquhoun HM, Lewis DF, Ben-Haida A, Hodge P (2003) Macromolecules 36:3775
230. Ben-Haida A, Hodge P, Colquhoun HM (2005) Macromolecules 38:722
231. Colquhoun HM, Dudman CC, Thomas M, O'Mahoney CA, Williams DJ (1990) J Chem Soc Chem Commun, p 336
232. Colquhoun HM, Zhu ZX, Williams DJ (2001) Org Lett 3:4031
233. Colquhoun HM, Zhu Z, Dudman CC, O'Mahoney CA, Williams DJ, Drew MGB (2005) Macromolecules 38:10413

234. Colquhoun HM, Zhu Z, Dudman CC (2005) Macromolecules 38:10421
235. Colquhoun HM, Lewis DF, Fairman RA, Baxter I, Williams DJ (1997) J Mater Chem 7:1
236. Wang YF, Chan KP, Hay AS (1995) Macromolecules 28:6371
237. Wang YF, Hay AS (1996) Macromolecules 29:5050
238. Wang YF, Chan KP, Hay AS (1996) Macromolecules 29:3717
239. Ding Y, Hay AS (1996) Macromolecules 29:6386
240. Wang YF, Chan KP, Hay AS (1996) J Appl Polym Sci 59:831
241. Wang YF, Chan KP, Hay AS (1996) React Polym 30:205
242. Wang YF, Hay AS (1997) Macromolecules 30:182
243. Ding Y, Hay AS (1997) Macromolecules 30:2527
244. Tsuchida E, Miyatake K, Hay AS (1998) Macromolecules 31:6469
245. Meng YZ, Hay AS (1999) J Appl Polym Sci 74:3069
246. Meng YZ, Tjong SC, Hay AS (2001) Polymer 42:5215
247. Colquhoun HM, Zolotukhin MG, Zhu Z, Hodge P, Williams DJ (2004) Macromol Rapid Commun 25:808
248. Zolotukhin MG, Fomine S, Colquhoun HM, Zhu Z, Drew MGB, Olley RH, Fairman RA, Williams DJ (2004) Macromolecules 37:2041
249. Liang ZA, Meng YZ, Li L, Du XS, Hay AS (2004) Macromolecules 37:5837
250. Ding Y, Hay AS (1997) Polymer 38:2239
251. Hay AS, Williams FJ, Relles HN, Boulette BM (1984) J Macromol Sci Chem A21:1065
252. Miyatake K, Hlil AR, Hay AS (2001) Macromolecules 34:4288
253. Iimori H, Shibasaki Y, Ando S, Ueda M (2003) Macromol Symp 199:23
254. Kihara N, Komatsu S, Takata T, Endo T (1999) Macromolecules 32:4776
255. Nomura N, Tsurugi K, Okada M (1999) J Am Chem Soc 121:7268
256. Nomura N, Yoshida N, Tsurugi K, Aoi K (2003) Macromolecules 36:3007
257. Nomura N, Tsurugi K, Okada M (2001) Angew Chem Int Ed 40:1932
258. Nomura N, Tsurugi K, RajanBabu TV, Kondo T (2004) J Am Chem Soc 126:5354
259. Kimura K, Kohama S, Yamashita Y (2002) Macromolecules 35:7545
260. Kimura K, Kohama S, Yamashita Y (2003) Macromolecules 36:5045
261. Kimura K, Kohama S, Yamazaki S (2006) Polym J 10:1005
262. Zhao D, Moore JS (2002) J Am Chem Soc 124:9996
263. Zhao D, Moore JS (2003) Macromolecules 36:2712
264. Zhao D, Moore JS (2003) J Am Chem Soc 125:16294
265. Shea KJ, Walker JW, Zhu H, Paz M, Greaves J (1997) J Am Chem Soc 119:9049
266. Busch BB, Paz MM, Shea KJ, Staiger CL, Stoddard JM, Walker JR, Zhou XZ, Zhu H (2002) J Am Chem Soc 124:3636
267. Busch BB, Staiger CL, Stoddard JM, Shea KJ (2002) Macromolecules 35:8330
268. Shea KJ, Lee SY, Busch BB (1998) J Org Chem 63:5746
269. Shea KJ, Busch BB, Paz MM (1998) Angew Chem Int Ed 37:1391
270. Wagner CE, Shea KJ (2001) Org Lett 3:3063
271. Goddard JP, Lixon P, Gall TL, Mioskowski C (2003) J Am Chem Soc 125:9242
272. Mondière R, Goddard JP, Carrot G, Gall TL, Mioskowski C (2005) Macromolecules 38:663
273. Montague RA, Green JB, Matyjaszewski K (1995) J Macromol Sci Pure Appl Chem A32:1497
274. Honeyman CH, Manners I, Morrissey CT, Allcock HR (1995) J Am Chem Soc 117:7035
275. Allcock HR, Reeves SD, de Denus CR, Crane CA (2001) Macromolecules 34:748
276. Wang B (2005) Macromolecules 38:643

277. Montague RA, Matyjaszewski K (1990) J Am Chem Soc 112:6721
278. Matyjaszewski K, Moore MK, White ML (1993) Macromolecules 26:6741
279. White ML, Matyjaszewki K (1997) Macromol Chem Phys 198:665
280. Allcock HR, Reeves SD, Nelson JM, Crane CA (1997) Macromolecules 30:2213
281. Nelson JM, Primrose AP, Hartle TJ, Allcock HR (1998) Macromolecules 31:947
282. Allcock HR, Prange R, Hartle TJ (2001) Macromolecules 34:5470
283. Chang Y, Prange R, Allcock HR, Lee SC, Kim C (2002) Macromolecules 35:8556
284. Chang Y, Powell ES, Allcock HR, Par SM, Kim C (2003) Macromolecules 36:2568
285. Prange R, Reeves SD, Allcock HR (2000) Macromolecules 33:5763
286. Allcock HR, Powell ES, Maher AE, Prange RL, de Denus CR (2004) Macromolecules 37:3635
287. Allcock HR, Powell ES, Chang Y (2004) Macromolecules 37:7163
288. Allcock HR, Nelson JM, Prange R, Crane CA, de Denus CR (1999) Macromolecules 32:5736
289. Prange R, Allcock HR (1999) Macromolecules 32:6390
290. Allcock HR, Prange R (2001) Macromolecules 34:6858
291. Nelson JM, Allcock HR (1997) Macromolecules 30:1854
292. Allcock HR, de Denus CR, Prange R, Laredo WR (2001) Macromolecules 34:2757
293. Allcock HR, Powell ES, Maher AE, Berda EB (2004) Macromolecules 37:5824
294. Lenz RW, Handlovits CE, Smith HA (1962) J Polym Sci 58:351
295. Newton AB, Rose JB (1972) Polymer 13:465
296. Hibbert DB, Sandall JPB (1988) J Chem Soc Perkin Trans II 1739
297. Robello DR, Ulman A, Urankar EJ (1993) Macromolecules 26:6718
298. Risse W, Heitz W (1985) Makromol Chem 186:1835
299. Percec V, Wang JH (1991) J Polym Sci A Polym Chem 29:63
300. Yokozawa T, Asai T, Sugi R, Ishigooka S, Hiraoka S (2000) J Am Chem Soc 122:8313
301. Yokozawa T, Muroya D, Sugi R, Yokoyama A (2005) Macromol Rapid Commun 26:979
302. Shibasaki Y, Araki T, Okazaki M, Ueda M (2002) Polym J 34:261
303. Yokozawa T, Ogawa M, Sekino A, Sugi R, Yokoyama A (2002) J Am Chem Soc 124:15158
304. Yokozawa T, Ogawa M, Sekino A, Sugi R, Yokoyama A (2003) Macromol Symp 199:187
305. Ohshimizu K, Shibasaki Y, Komura M, Nakajima K, Ueda M (2007) Chem Lett 36:742
306. Ohshimizu K, Shibasaki Y, Ueda M (2007) Polym J 39:777
307. Izawa Y, Hiraoka S, Yokozawa T (2001) Polym Prepr Jpn 50:215
308. Sugi R, Hitaka Y, Sekino A, Yokoyama A, Yokozawa T (2003) J Polym Sci A Polym Chem 41:1341
309. Sugi R, Yokoyama A, Yokozawa T (2003) Macromol Rapid Commun 24:1085
310. Kim S, Kakuda Y, Yokoyama A, Yokozawa T (2007) J Polym Sci A Polym Chem 45:3129
311. Sugi R, Hitaka Y, Yokoyama A, Yokozawa T (2005) Macromolecules 38:5526
312. Yokoyama A, Hirabayashi K, Aiyama A, Yokozawa T (2003) Macromolecules 36:4328
313. Iwashita K, Yokoyama A, Yokozawa T (2005) J Polym Sci A Polym Chem 43:4109
314. Yokozawa T, Suzuki Y, Hiraoka S (2001) J Am Chem Soc 123:9902
315. Suzuki Y, Hiraoka S, Yokoyama A, Yokozawa T (2004) J Polym Sci A Polym Chem 42:1198
316. Suzuki Y, Hiraoka S, Yokoyama A, Yokozawa T (2003) Macromolecules 36:4756
317. Ajioka N, Suzuki Y, Yokoyama A, Yokozawa T (2007) Macromolecules 40:5294
318. Yokozawa T, Taniguchi T, Suzuki Y, Yokoyama A (2002) J Polym Sci A Polym Chem 40:3460

319. Shibasaki Y, Araki T, Nagahata R, Ueda M (2005) Eur Polym J 41:2428
320. Kovalev A, Takeuchi K, Barzykin A, Asai M, Ueda M, Rusanov A (2005) Macromol Chem Phys 206:2112
321. Sugi R, Yokoyama A, Furuyama T, Uchiyama M, Yokozawa T (2005) J Am Chem Soc 127:10172
322. Ohishi T, Sugi R, Yokoyama A, Yokozawa T (2006) J Polym Sci A Polym Chem 44:4990
323. Sugi R, Ohishi T, Yokoyama A, Yokozawa T (2006) Macromol Rapid Commun 27:716
324. Yokoyama A, Miyakoshi R, Yokozawa T (2004) Macromolecules 37:1169
325. Iovu MC, Sheina EE, Gil RR, McCullough RD (2005) Macromolecules 38:8649
326. Sheina EE, Liu J, Iovu MC, Laird DW, McCullough RD (2004) Macromolecules 37:3526
327. Miyakoshi R, Yokoyama A, Yokozawa T (2004) Macromol Rapid Commun 25:1663
328. Miyakoshi R, Yokoyama A, Yokozawa T (2005) J Am Chem Soc 127:17542
329. Jeffries-El M, Sauvé G, McCullough RD (2005) Macromolecules 38:10346
330. McCullough RD (1998) Adv Mater 10:93
331. Adachi I, Miyakoshi R, Yokoyama A, Yokozawa T (2006) Macromolecules 39:7793
332. Vallat P, Lamps JP, Schosseler F, Rawiso M, Catala JM (2007) Macromolecules 40:2600
333. Yokozawa T, Adachi I, Miyakoshi R, Yokoyama A (2007) High Perform Polym 19:684
334. Liu J, Sheina E, Kowalewski T, McCullough RD (2002) Angew Chem Int Ed 41:329
335. Iovu MC, Jeffries-EL M, Sheina EE, Cooper JR, McCullough RD (2005) Polymer 46:8582
336. Dai CA, Yen WC, Lee YH, Ho CC, Su WF (2007) J Am Chem Soc 129:11036
337. Radano CP, Scherman OA, Stingelin-Stutzmann N, Müller C, Breiby DW, Smith P, Janssen RAJ, Meijer EW (2005) J Am Chem Soc 127:12502
338. Miyakoshi R, Shimono K, Yokoyama A, Yokozawa T (2006) J Am Chem Soc 128:16012
339. Yokoyama A, Suzuki H, Kubota Y, Ohuchi K, Higashimura H, Yokozawa T (2007) J Am Chem Soc 129:7236
340. Sinclair DJ, Sherburn MS (2005) J Org Chem 70:3730
341. Dong CG, Hu QS (2005) J Am Chem Soc 127:10006
342. Weber SK, Galbrecht F, Scherf U (2006) Org Lett 8:4039
343. Yokozawa T, Suzuki H (1999) J Am Chem Soc 121:11573
344. Yokozawa T, Maeda D, Hiyama N, Hiraoka S (2002) Macromol Chem Phys 51:1265

Editor: Shiro Kobayashi

Adv Polym Sci (2008) 217: 79–119
DOI 10.1007/12_2008_136

Published online: 6 March 2008

Polymer Hybrids Based on Polyolefins – Syntheses, Structures, and Properties

Nobuo Kawahara (✉) · Junji Saito · Shingo Matsuo · Hideyuki Kaneko · Tomoaki Matsugi · Norio Kashiwa

Research Center, Mitsui Chemicals, Inc., 580-32 Nagaura, Sodegaura-City, 299-0265 Chiba, Japan
Nobuo.Kawahara@mitsui-chem.co.jp

Abstract Recent advances in polyolefin chemistry have led to the creation of polymer hybrid linking between different polymer segments to broaden their applications towards use in highly profitable fields. Synthesized polymer hybrids based on polyolefins are expected to possess the properties of each polymer segment. Several methodologies have been reported for the preparation of polyolefin hybrids which employ processes, such as radical, anionic, cationic polymerizations, and post-polymerization reactions, bringing certain improved and unique properties to ordinary polyolefins. These new polymer hybrids can be synthesized using functionalized polyolefins, such as polyolefin (PO) macroinitiator, PO macromonomer, and reactive PO. Synthesized polymer hybrids based on PO, with a well-defined molecular structure, show excellent properties compared to conventional POs, including compatibility and mechanical strength.

Keywords Block copolymer · Controlled-nano architecture · Functionalization · Polymer hybrid · Polyolefin

Abbreviations

acac	Acetylacetonate
ADMS	Allyldimethylsilane
AFM	Atomic force microscopy
AIBN	2,2′-Azobisisobutyronitrile
ATR/IR	Attenuated total reflection infrared
ATRP	Atom transfer radical polymerization
9-BBN	9-Borabicyclononane
BzInd	Benzindenyl
n-Bu	*n*-Butyl
s-Bu	*sec*-butyl
t-Bu	*tert*-butyl
ε-CL	ε-Caprolactone
Cp	Cyclopentadienyl
CRP	Controlled radical polymerization
CTA	Chain transfer agent
DSC	Differential scanning calorimeter
EPR	Poly(ethylene-*co*-propylene)
EBR	Poly(ethylene-*co*-butene)
EPDM	Ethylene-propylene terpolymer
Et	Ethyl, ethylene
FEP	Tetrafluoroethylene-hexafluoropropylene copolymer
GPC	Gel permeation chromatography
HEMA	2-Hydroxyl-ethylmethacrylate
Ind	Indenyl
IR	Infrared
MA	Methacrylate
MAH	Maleic anhydride
MAO	Methylaluminoxane
Me	Methyl
Mw	Weight average molecular weight
Mn	Number average molecular weight
p-MS	*p*-Methylstyrene
NBS	*N*-bromosuccinimide
NMP	Nitroxide mediated polymerization
NMR	Nuclear magnetic resonance
Oct	Octyl
P4MP1	Poly(4-methylpent-1-ene)
PBMA	Poly(butyl methacrylate)
PnBA	Poly(*n*-butyl acrylate)
PtBA	Poly(*tert*-butyl acrylate)
PCL	Polycaprolactone
PE	Polyethylene
PEO	Poly(ethylene oxide)
PMDETA	*N*,*N*,*N*′,*N*″,*N*″-pentamethyldiethylenetriamine
PMMA	Poly(methyl methacrylate)
PO	Polyolefin
PP	Polypropylene
sPP	*Syndiotactic*-polypropylene
py-PP	Pyrolysis polypropylene

PPG	Poly(propylene oxide)
PS	Polystyrene
Ph	Phenyl
RAFT	Reversible addition fragmentation chain transfer
SEM	Scanning electron microscopy
TEM	Transmission electron microscopy
TEMPO	2,2,6,6-tetramethyl-1-piperidinyloxy
THF	Tetrahydrofurane
TMEDA	*N,N,N′,N′*-tetramethylethylenediamine
THInd	Tertahydroindenyl
T_g	Glass transition temperature

1 Introduction

Polyolefins (POs), represented by polyethylene (PE) and polypropylene (PP), are indispensable materials with a certain social impact and are the most widely used commercial polymers with a worldwide production of more than one thousand million tons per year, accounting for 50% of the total amounts of synthetic polymers. They are used in our daily lives for such applications as films, pipes, automobile parts, electric and electronic parts, medical and sanitary products, toys, construction materials, and agricultural materials, which is to say, they are used in various fields from the cradle to the grave. This is due to POs possessing excellent properties, such as high mechanical strength, flexibility, chemical stability, and processability. POs are also responsible for catalyst and process innovations since the discovery of the Ziegler-Natta catalyst systems in 1950s [1]. In addition, polyolefins are clean materials, because they are composed of only carbon and hydrogen atoms. Also, they are easily recycled. These advantages have led to a reduction in environmental pollution.

Despite the great success of polyolefins, deficiencies, such as lack of compatibility with other materials including polymers, have limited some of their end-uses, particularly regarding adhesion, dyeability, paintability, and printability. Therefore, many research activities have been geared toward the preparation of functional polyolefins. In addition, many scientists have tried to overcome these disadvantages by introducing suitable functional groups to polyolefins, thereby broadening the applications of polyolefins towards highly profitable fields [2].

So far, theoretically, there have been two possible approaches for the synthesis of functional polyolefins, namely, (a) direct copolymerization of olefins with functional monomers using polymerization and catalyst technologies, and (b) post-polymerization reaction with polyolefins.

The first approach has found limited applications. Concerning the direct copolymerization of olefins with functional monomers, studies of copolymerization of olefins by metallocene catalysts have been reported with functional monomers, such as ω-chloro-α-olefins [3], 5-(*N*,*N*-diimidopropylamino)-1-pentene [4], cubic silsequioxane [5,6], 6-*tert*-butyl-2-(1,1-dimethylhept-6-enyl)-4-methyl phenol [7,8], hindered amine light stabilizers (HALS) [9], carbazole-based comonomer [10], Oxazoline-based comonomer [11], allyl-anisole [12], hydroxyl olefins, etc.

The second synthetic route has found widespread application in PO technology. Regarding the post-polymerization reaction, it is common for polyolefins to be treated with a radical initiator and functional chemicals, such as maleic anhydride, halogen, (meth)acrylate monomers, and so on. Chemical modification of the pre-formed polyolefins has usually been carried out in situ during the fabrication process to reduce the production cost. In addition, surface modification of polymer materials continues to be an important area, because of the many cases where the surface properties of polymers affect their utilities. Several methods, such as radical modification with maleic anhydride [13,14], oxidation of the surface of PE film [15,16], and plasma treatment of polymer films [17,18], were good for the improvement of hydrophilicity of polyolefin materials. However, there are no facile reaction sites in saturated polyolefins. The only way is to activate the polymer by breaking some stable C–H bonds and forming free radicals along the polymer chain, as mentioned above. These functionalization reactions are usually accompanied by many undesirable side reactions, such as crosslinking and degradation. Unsaturated groups of polyolefins were converted by several functional groups, such as the epoxy group, silyl group, hydroxyl group, maleic anhydride group, etc. [19]. Unsaturated structures, such as vinyl and vinylidene, are generally observed in pyrolized polyolefins. Metallocene catalyst systems, however, offer significant advantages for the synthesis of functional polyolefins, because they are able to produce polyolefins possessing unsaturated structures at the end of (or in the) polymer chain. The chain transfer reactions in metallocene catalyzed polymerization have led to terminal double bonds that are used to further chemical reactions. Halogenation [20,21] and amination [22,23] of terminally unsaturated groups of PP have been described in previous papers. Oxazoline terminated PP [24] was synthesized from metallocene catalyzed PP having a terminally vinylidene group. Terminally unsaturated polyolefins, possessing vinyl or vinylidene groups, are valuable as macromonomers for cationic, radical, and coordination polymerization reactions [25,26].

Recent advances in polyolefin chemistry have led to the creation of polymer hybrids linking between different polymer segments, namely crystal polyolefins, amorphous polyolefens, and other polar polymers (Fig. 1) to create polyolefin materials with novel functionalities, including polarity, etc.

Synthesized polymer hybrids based on polyolefins are expected to possess the properties of each polymer segment. Several methodologies for the preparation of polyolefin hybrids have been reported in recent papers [27,28]. In particular, processes such as radical, anionic, cationic polymerizations and

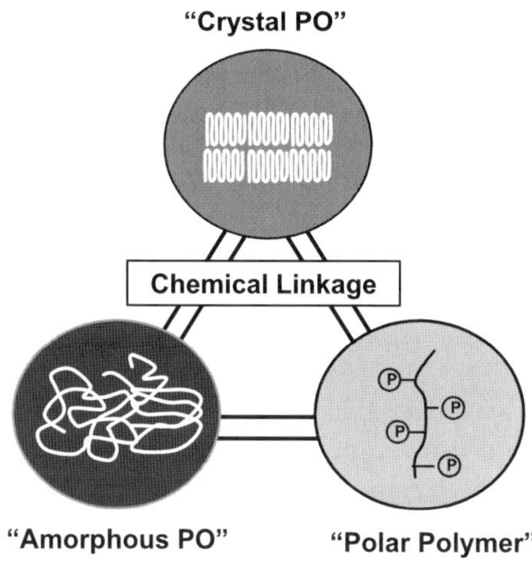

Fig. 1 Concept of polymer hybrid

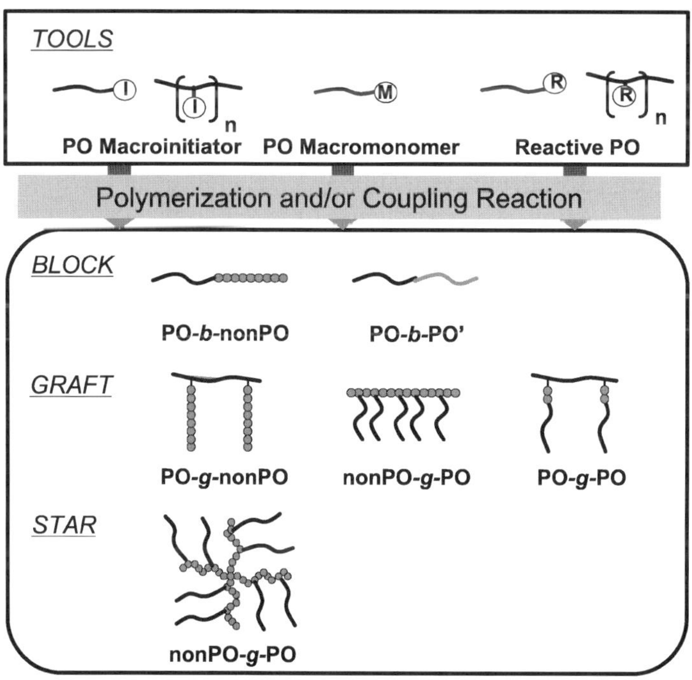

Fig. 2 Three tools for preparing polymer hybrids based on POs with well-defined molecular structures

post-polymerization reactions have resulted in ordinary polyolefins with improved and unique properties that broaden their applications. So far, Chung's group has proposed the synthetic methods of polymer hybrids using polymerization and post-polymerization techniques. Other research groups have also reported on synthetic procedures for polyolefin hybrids using functionalized polyolefins. We herein propose three tools to produce polymer hybrids based on polyolefins using PO macroinitiator, PO macromonomer and reactive PO. These methodologies offer a comprehensive coverage of polyolefin block or graft copolymers with well-defined molecular structures, such as linear block type, graft type, and multi-armed type copolymers (Fig. 2). It is possible to design a new class of PO-based polymers with unique topologies and compositions. Synthesized polymer hybrids based on POs have different characteristics resulting in new functions as new materials as a result of their synergic effect.

In this article, we would like to review recent advances in polymer hybrids based on polyolefins and to classify them into four methodologies, namely, PO macroinitiator, PO macromonomer, reactive PO and living copolymerization of olefins, with their well-defined structures and applications.

2 Synthetic Methodologies of Polyolefin Hybrids

2.1 Polymerization with Polyolefin Macroinitiator

The concept of PO macroinitiators centers on the introduction of an initiation moiety into an olefinic polymer chain for polymerization. The most effective route for preparing PO macroinitiators is by employing functional polyolefins containing hydroxyl groups or other reactive groups. These functional POs are prepared by copolymerization of olefins with functional monomers and post-polymerization reaction, as mentioned above. In the case where an initiation moiety was at the chain-end of the polyolefins, a block type copolymer is produced. It has been reported that thiol-terminated PP was used as polymeric chain transfer agent in styrene and styrene/acrylonitrile polymerization to form polypropylene-*block*-polystyrene (PP-*b*-PS) and polypropylene-*block*-poly(styrene-co-acrylonitrile) (PP-*b*-SAN) block copolymer [19]. On the other hand, polymer hybrids with block and graft structures can be produced if initiation moieties are in the polymer chain.

2.1.1
Living Anionic Polymerization with PO Macroinitiator

For Block Copolymers:

In the 1980s, a synthetic method to produce AB block copolymers of propylene and tetrahydrofurane (THF) was proposed [29]. Polypropylene-*block*-poly(THF) was prepared by a combination of living polymerization of propylene with a $V(acac)_3$ catalyst and the living polymerization of THF. Its synthesis was based on the transformation of living polypropylene chain ends to cationic ones, which initiated the living polymerization of THF.

The methodology to synthesize polymer hybrids by living anionic polymerization is shown in Fig. 3 [30]. Polyolefins containing *p*-tolyl groups have been used to initiate anionic polymerization by the lithiation of the methyl moiety using alkyl lithium and amine compounds system.

As methods to synthesize block copolymers, they have reported on the preparation of *p*-tolyl group terminated polyolefins. Copolymerizations of α-olefin with *p*-methylstyrene (*p*-MS) and hydrogen simultaneously in the presence of metallocene, which was $Me_2Si(4\text{-}Ph\text{-}2\text{-}Me(Ind))_2ZrCl_2$, can produce *p*-MS terminated PO. In this case, during the polymerization of propylene, the propagation Zr–C site would react with *p*-MS to form *p*-MS terminated PP. The catalytic Zr–C site in the compound became inactive to both propylene and *p*-MS, due to the combination of steric hindrance between the active site (Zr–C) and incoming monomer, which is a dormant site [31]. With the presence of hydrogen as chain transfer agent, the dormant Zr–C site could react with hydrogen to form *p*-MS terminated PP and regenerate the Zr–H site, which is capable of reinitiating the polymerization of propylene and *p*-MS. The lithiation of the *p*-tolyl group at the end of the PO chain with sec-butyl lithium (*s*-BuLi)/*N*,*N*,*N*′,*N*′-tetramethylethylenediamine (TMEDA) gives PO macroinitiator for a living anionic polymerization.

Block copolymers, polyethylene-*block*-polystyrene (PE-*b*-PS) and PP-*b*-PS, were prepared by the polymerization of styrene with terminally lithiated PO macroinitiators [31].

It was also reported that terminally borane-containing POs is available as another macroinitiator to prepare block copolymers (Fig. 4). These polymers were prepared by (1) the metallocene-catalyzed (co)polymerization of olefin(s) with organoborane compounds, for example, 9-borabicyclononane (9-BBN), as chain transfer agents [32], or by (2) the hydroboration of terminally unsaturated polyolefins with BBN [33–36].

In case of (1), they proposed the use of dialkylborane compounds as chain transfer for catalytic polymerization. These compounds form stable dimers in organic solvents, such as toluene or hexane. The B–H groups in these boranes were very reactive toward a metal-carbon (M–C) active site formed during the catalytic polymerization with metallocenes. The transfer reaction

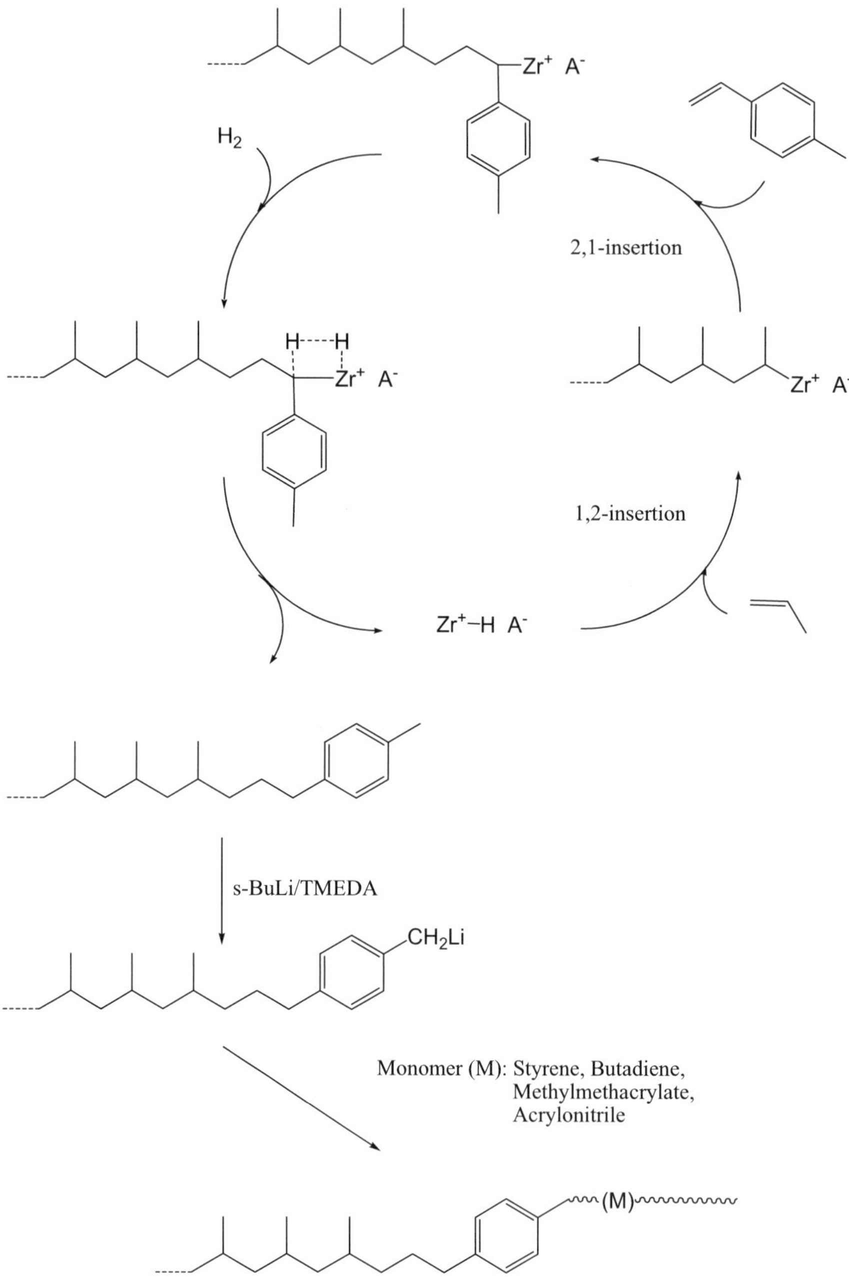

Fig. 3 Synthetic route of PP hybrids via terminally lithiated PP macroinitiator

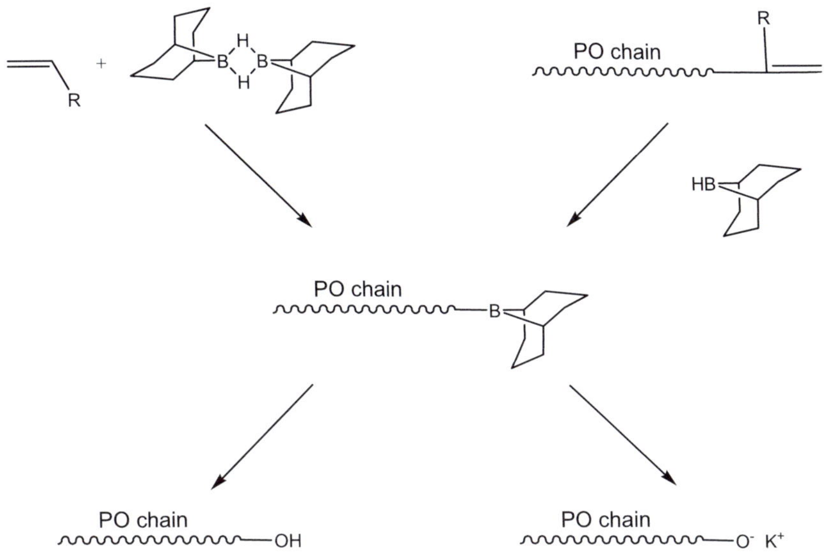

Fig. 4 Synthesis of PO macroinitiator via terminally borane-containing POs for living anionic polymerization

between B–H and the M–C active site results in a polyolefin terminated with a 9-BBN group and in the regeneration of the active species M–H. The molar mass of the polymer formed was inversely proportional to the concentration of the borane-based compounds, since the transfer reaction was a dominant process in this polymerization mechanism. In case of (2), borane-containing polyolefins can also be prepared by hydroboration of terminally unsaturated polyolefins prepared by metallocene-catalyzed olefin polymerization, with BBN compounds.

Introduced as a chain-end group, the BBN group was converted to an alkoxide group, that is, $-O^-K^+$, an efficient initiator for anionic polymerization for ethylene oxide. The ring opening living anionic polymerizations of ethylene oxide were carried out with potassium alkoxide polyolefins to prepare polyethylene-*block*-poly(ethylene oxide) (PE-*b*-PEO), poly(ethylene-*co*-styrene)-*block*-poly(ethylene oxide), and poly(ethylene-*co*-octene)-*block*-poly(ethylene oxide) (EOR-*b*-PEO) [37].

Polyethylene-*block*-polycaprolactone (PE-*b*-PCL) was produced with terminally hydroxylated polyolefins prepared from terminally borane-containing polyolefins [38]. The terminal borane group was converted to a hydroxyl group by an oxidation reaction with $NaOH/H_2O_2$ reagent, and to an active initiator by metallation reaction of the terminal OH group with diethylaluminum chloride. This was adopted in ring-opening anionic polymerization of ε-caprolactone to form the resulting block copolymer.

The synthetic procedure of PE-*b*-PCL using hydroxyl terminated polyethylene was reported [39]. Terminally hydroxylated polyethylene was prepared during a metallocene-catalyzed polymerization using controlled chain transfer reaction with alkylaluminum compounds. PE-*b*-PCL block copolymer was synthesized from terminally hydroxylated PE and ε-caprolactone (ε-CL) using $Sn(Oct)_2$ as a catalyst for ring opening polymerization.

For Graft Copolymers:

To synthesize graft copolymers, it has also been reported on the preparation of *p*-tolyl group contained polyolefins. In general, the preparation of *p*-MS containing PO copolymers has been greatly enhanced by the metallocene catalysts with constrained ligand geometry, having a spatially opened catalytic site, allowing the effective incorporation of styrenic comonomers, which was quite difficult using Ziegler-Natta catalyst systems. In the copolymerization reaction, *p*-MS comonomers exhibit good incorporation with ethylene using metallocene catalysts, such as $[C_5Me_4(SiMe_2N\text{-}t\text{-}Bu)]TiCl_2$, Cp_2ZrCl_2 or $Et(Ind)_2ZrCl_2$ activated with methylaluminoxane (MAO) [40–42]. The resulting copolymer contained 40 mol% of *p*-MS. Both DSC and ^{13}C nuclear magnetic resonance (NMR) results showed the random distribution of *p*-MS units in the copolymer chain. As a result, methyl groups of *p*-MS units in PE can be effectively lithiated by a *s*-BuLi and TMEDA system at ambient temperature to prepare the PO macroinitiator. Graft copolymers, polyethylene-*graft*-polystyrene (PE-*g*-PS), polyethylene-*graft*-poly(methyl methacrylate) (PE-*g*-PMMA), and polyethylene-*graft*-polyacrylonitrile were prepared from these PO macroinitiators in cyclohexane by living anionic polymerization [42–44]. Also, graft copolymers, PE-*g*-PS, poly(ethylene-*co*-propylene)-*graft*-polystyrene (EPR-*g*-PS), and poly(ethylene-co-octene)-*graft*-polystyrene (EOR-*g*-PS) were obtained [45, 46].

Olefin copolymers containing a hydroxyl group can be used as the PO macroinitiator for ring opening anionic polymerization. Copolymerization of olefins with hydroxyl olefin monomers protected with alkylaluminum compounds by metallocene/MAO catalyst systems gave hydroxylated olefin copolymers, such as terminally hydroxylated PE, hydroxylated PE, and hydroxylated PP [47–49]. The control of hydroxylated PE structures was investigated by Kashiwa et al. in the copolymerization of ethylene with allyl alcohol or 10-undecen-1-ol with a specific metallocene, MAO and a trialkylaluminum catalyst system through changes in the copolymerization conditions [50]. The incorporation of allyl alcohol into the PE backbones was controllable through changes in the trialkylaluminum, leading to terminally hydroxylated PE or a copolymer possessing hydroxyalkyl side chains. The copolymerization of ethylene with 10-undecen-1-ol gave copolymers with hydroxyalkyl side chains of various contents with a variety of molecular weights through changes in the copolymerization conditions. The obtained hydroxylated PE could be used as PO macroinitiators, which allowed polar polymer segments to grow on the PE backbones, leading to the creation of graft copolymers that possessed PE and polar polymer segments (Fig. 5).

Graft copolymer, polyethylene-*graft*-poly(propylene oxide) (PE-*g*-PPG), has been synthesized by ring opening anionic polymerization of propylene oxide with a phosphazene catalyst and hydroxylated polyethylene (M_n = 12 400, [OH] = 5 units/chain). Polymerization of propylene oxide was carried out in tetraline at 120 °C for 20 hours. The ^{13}C NMR analysis of PE-*g*-PPG suggested that all the hydroxyl groups were consumed for propylene oxide polymerization (Fig. 6).

Polyethylene-*graft*-polycaprolactone (PE-*g*-PCL) was synthesized by ring opening anionic polymerization of ε-CL with a triethylalminum catalyst and hydroxylated PE. Polymerization of ε-CL was performed in toluene for 24 hours. The obtained PE-*g*-PCL, in which 36–80 mol % of the hydroxyl groups worked as initiators for ε-CL polymerization, possessed 29 units of ε-CL in polymer hybrids (Fig. 7).

Microstructures of (1) PE-*g*-PPG polymer hybrid and (2) the blended sample of PE and PPG were observed by transmission electron microscopy (TEM) images after the preparation of press sheets of each polymer sample at 200 °C. The TEM images of the resulting polymer hybrid reveal the nanometer level microphase-separation morphology between the PE segment and the PPG segment compared with the PE/PPG blended polymer. From the result, the nanodispersion of different segments in polymer hybrids is possible, but different from the blended polymer sample (Fig. 8).

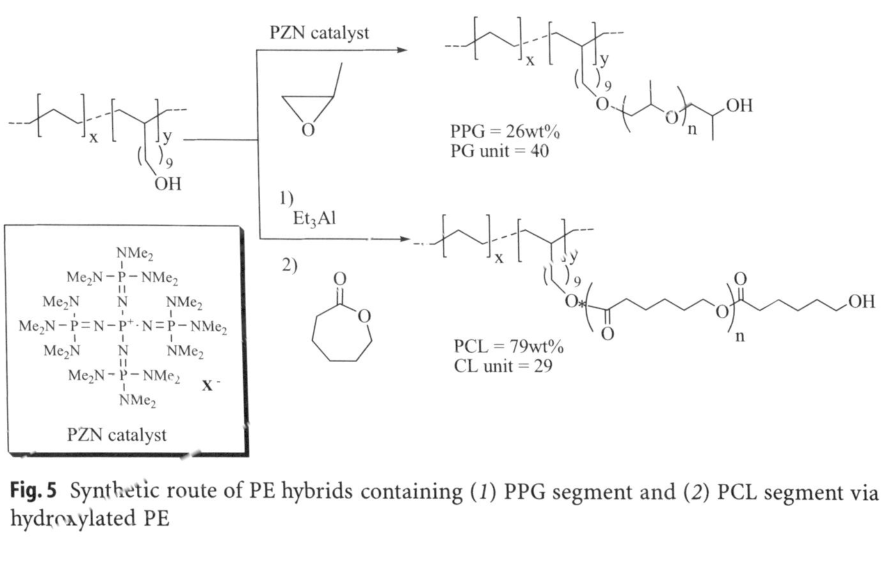

Fig. 5 Synthetic route of PE hybrids containing (*1*) PPG segment and (*2*) PCL segment via hydroxylated PE

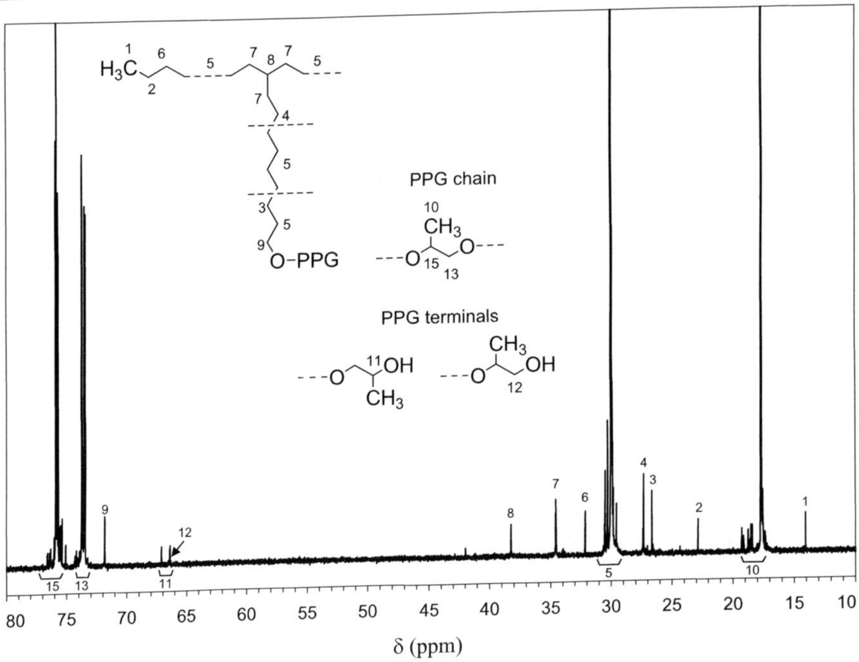

Fig. 6 ^{13}C NMR spectrum of PE-*g*-PPG

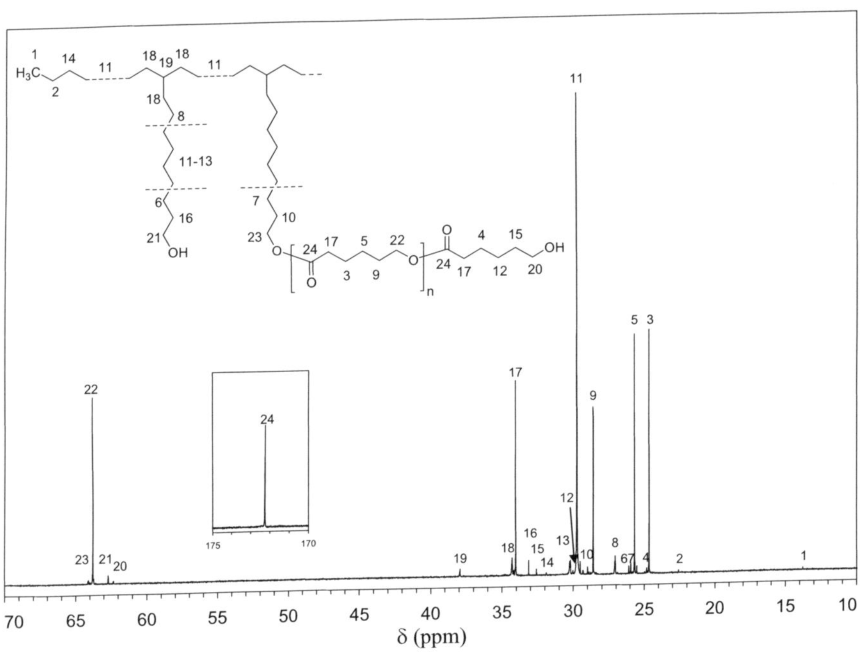

Fig. 7 ^{13}C NMR spectrum of PE-*g*-PCL

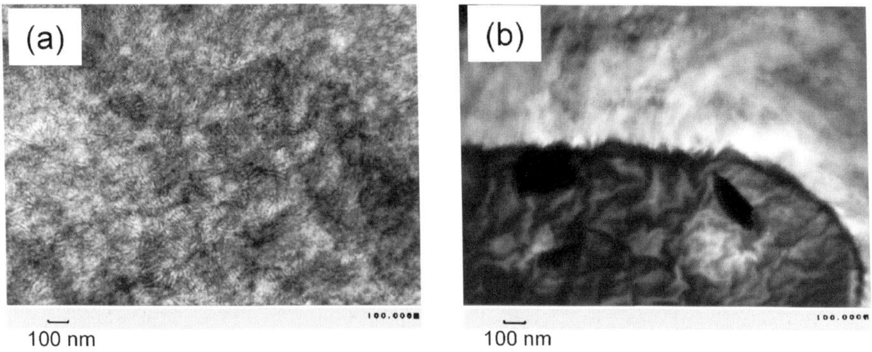

Fig. 8 TEM images of **a** PP-*g*-PPG and **b** homo-PE/homo-PPG blended sample (homo-PE/homo-PPG = 6/4 wt ratio)

It is known that borane containing polymers were converted to hydroxylated polyomers, as mentioned above. From borane-containing PE, PE-*g*-PCL was also synthesized [51].

2.1.2
Living Radical Polymerization Via Borane Groups

Controlled radical polymerization (CRP) is an attractive tool, because of the resultant controllability of polymerization, and because of it being a versatile method to synthesize of well-defined polymer hybrids. The three main radical polymerization techniques, ATRP, NMP, and RAFT polymerization, have thus been employed. Other techniques, such as the oxidation of borane groups, have also been studied. In general, using CRP techniques, block copolymers can be synthesized from terminally functionalized PO as PO macroinitiator, and block copolymers can be prepared from functionalized PO produced by the copolymerization of olefins with functional monomers.

The living radical polymerization via borane side groups has been proposed to synthesize polymer hybrids (Fig. 9). Copolymerization of olefins with the borane monomer, 5-hexenyl-9-borabicyclononane (5-hexenyl-9-BBN), was deemed a very convenient way to prepare borane-containing polyolefins [52, 53]. Alternatively, borane-containing POs were prepared by hydroboration of unsaturated POs, such as poly(propylene-*co*-1,4-hexadiene) and poly(ethylene-*co*-propylene-co-1,4-hexadiene) [53]. These POs containing borane side groups synthesized by metallocene catalyst systems could be oxidized to produce a peroxyborane (C–O–O–B) moiety. The peroxyborane group would decompose by itself, followed by a homolytical cleavage of peroxide to generate an alkoxy radical (C–O*) and a borinate radical (*O–B) [54]. The alkoxy radical moiety (C–O*) would be very reactive and could be used as an initiator of radical polymerization. This radical polymerization was carried out as living nature.

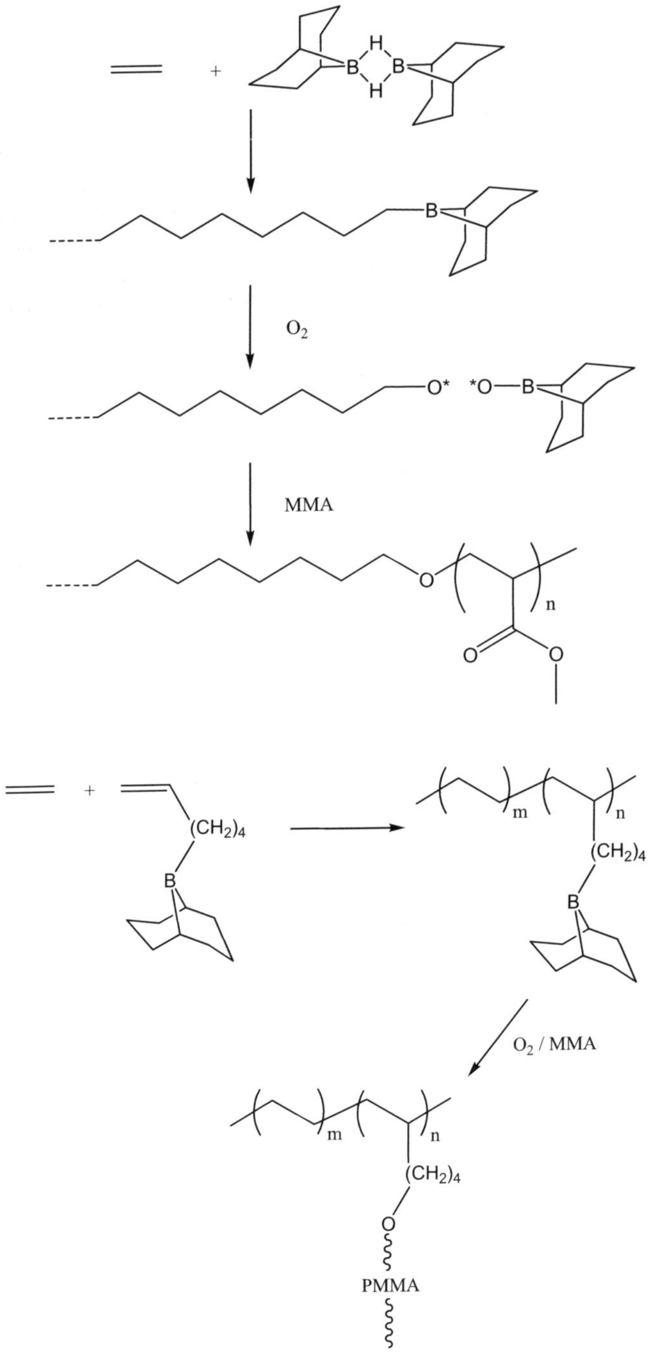

Fig. 9 Synthetic route of PE hybrids by living radical polymerization via borane-containing PE

For Block Copolymers:

Block copolymers can be produced from terminally borane-containing polyolefins. These borane-containing POs can be synthesized by the metallocene-catalyzed (co)polymerization of olefin(s) monomer with 9-BBN as a chain transfer agent or by the metallocene catalyzed copolymerization of olefins with allyl-9-BBN [55, 56], as referred to above. Alternatively, borane-containing POs were prepared by hydroboration of terminally unsaturated PO, for instance, terminally vinyl PE and terminally vinylidene PP [33–35, 57]. Such method could produce diblock copolymers, such as polyethylene-*block*-poly(methyl methacrylate) (PE-*b*-PMMA), polypropylene-*block*-poly(methyl methacrylate) (PP-*b*-PMMA), polypropylene-*block*-poly(butyl methacrylate) (PP-*b*-PBMA), and PP-*b*-PS.

For Graft Copolymers:

Graft copolymers can be synthesized from borane-containing POs. POs having borane groups can also be prepared by copolymerization of olefins with ω-borane-α-olefins, such as 5-hexenyl-9-BBN [58–62]. Copolymers of 5-hexenyl-9-BBN and propylene, or other α-olefins were obtained using $TiCl_3/Al$ compound catalyst systems. Copolymerization of ethylene with 5-hexenyl-9-BBN was carried out with $Et(Ind)_2ZrCl_2$ or Cp_2ZrCl_2 activated with MAO.

In addition, borane-containing POs can be prepared by copolymerization of olefin with borane monomers or by hydroboration of polyolefins including unsaturated groups, such as olefin-divinylbenzene copolymer and olefin-diene copolymers. Many kinds of graft copolymers, such as polyethylene-*graft*-poly(vinyl alcohol), PE-*g*-PMMA, polypropylene-*graft*-poly-(maleicanhydride-*co*-styrene), polypropylene-*graft*-poly(methacrylic acid), polypropylene-*graft*-poly(vinyl alcohol), polypropylene-*graft*-polycaprolactone (PP-*g*-PCL), polypropylene-*graft*-poly(methyl methacrylate) (PP-*g*-PMMA), poly(ethylene-*co*-propylene)-*graft*-poly(methyl methacrylate) (EPR-*g*-PMMA), and poly(ethylene-*co*-propylene)-*graft*-poly(maleic anhydride-*co*-styrene), have been synthesized by such a method resulting in controllable composition and molecular microstructures [63–66].

2.1.3
Atom Transfer Radical Polymerization

Among the several controlled radical polymerization methods, atom transfer radical polymerization (ATRP) is one of the most attractive techniques. The proposed mechanism of ATRP is shown in Fig. 10. The radicals were formed by atom transfer of a halogen from a dormant species to a metal. This transfer is possible due to the capacity of the metal to change its oxidation state. This

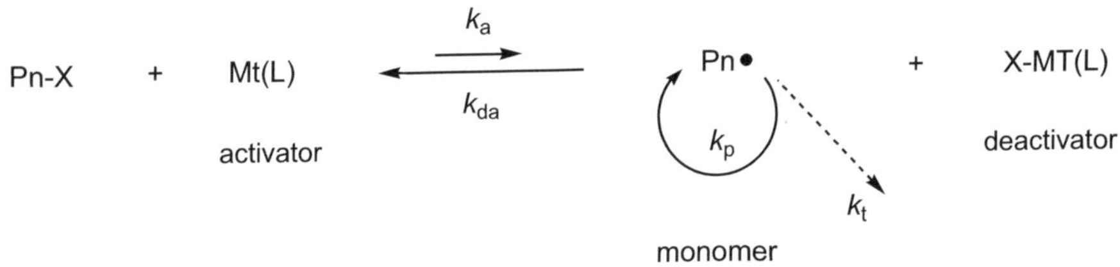

Fig. 10 ATRP mechanism

reversible reaction allows the establishment of equilibrium between dormant and active species, ensuring controlled polymerization.

For Block Copolymers:

To prepare block copolymers by ATRP, the initiation site for living radical polymerization can be introduced at the end of a polymer chain. In this context, terminally functionalized POs are useful for the synthesis of block copolymers.

It has been reported by Matsugi et al. that PE-*b*-PMMA can be synthesized by ATRP techniques with a CuBr and an *N*,*N*,*N*′,*N*″,*N*″-pentamethyldiethylenetriamine (PMDETA) catalyst system, or with a $RuCl_2(PPh_3)_3$ and dibutylamine catalyst system. They have used terminally hydroxylated PE, produced by the metallocene-catalyzed polymerization of ethylene with allyl alcohol as starting material. The PE macroinitiator was converted to terminally brominated PE by the bromination of obtained PE–OH with bromo-isobutyric bromide (Fig. 11) [67].

The resulting PE-*b*-PMMA was purified by soxhlet extraction with THF and characterized by NMR, DSC, and TEM micrography (Table 1). The TEM of the obtained PE-*b*-PMMA revealed unique morphological features which depended on the content of the PMMA segment. The block copolymer possessing 75 wt % PMMA contained 50–100 nm spherical polyethylene lamellae uniformly dispersed in the PMMA matrix (Fig. 12). Moreover, the PE-*b*-PMMA block copolymers effectively compatibilized homo-PE and homo-PMMA at a nanometer level (Fig. 13).

Terminally brominated PE as PE macroinitiator can be produced by other methods. It has been reported that vinyl terminated PE produced by a bis(phenoxy-imine)metal complex and MAO catalyst system (M_n = 1800, M_w/M_n = 1.70) was converted to terminally 2-bromoisobutyrate PE through the addition reaction of 2-bromoisobutyric acid to the vinyl chain end. Polyethylene-*block*-poly(*n*-butyl acrylate) (PE-*b*-PnBA) from terminally brominated PE by ATRP procedure has also been produced [68]. It was reported that degenerative transfer coordination polymerization with an iron complex can be used to prepare terminally brominated PE as a macroinitiator [69]. A Zn-terminated PE prepared using an iron complex and diethylzinc,

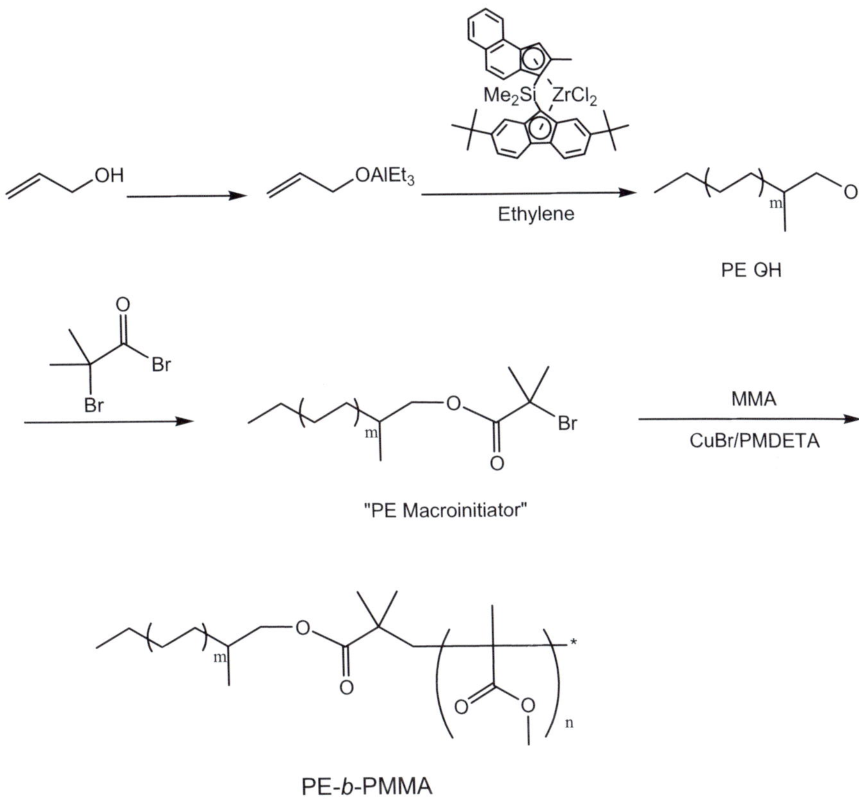

Fig. 11 Synthetic route of PE-*b*-PMMA by ATRP process

Table 1 Preparation results of polyolefin hybrids, PE-*b*-PMMA, by ATRP [a]

Entry	Catalyst	PE-*t*-Br (mM[Br])	[MMA]/[PE-Br] (Molar ratio)	Polymn. Time (h)	Yield (g)	MMA Content (wt%) [b]	M_w [c] (g/mol)	T_m [d] (°C)
1	CuBr	0.41g (4.0)	600	6.0	0.60	27	31 700	132.3
2	CuBr	0.82g (8.0)	600	6.0	0.91	55	–	130.2
4	CuBr	0.49g (11.0)	400	7.0	1.14	60	40 700	129.9
5	CuBr	0.49g (11.0)	600	7.0	1.95	75	49 400	129.2
Ref.		(PE-*t*-OH)					26 500	132.5

[a] Polymerization conditions; [PE – Br]/[CuBr]/[PMDETA] = 1/1/2 (molar ratio), temperature: 120 °C, solvent: *o*-xylene

[b] Determined by ^{1}H NMR

[c] Determined by GPC and calibrated with PE

[d] Measured by DSC

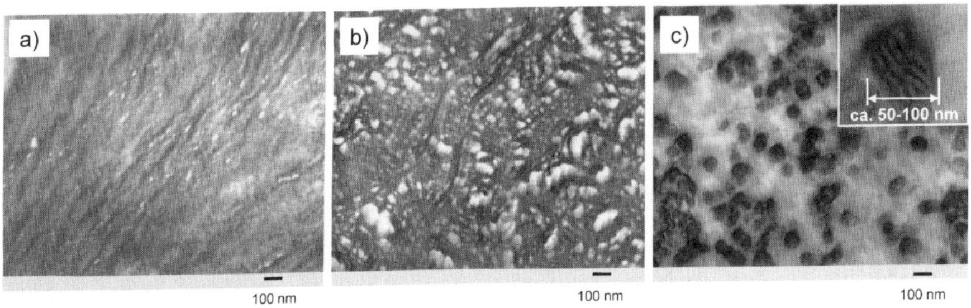

Fig. 12 TEM images of PE-*b*-PMMA with PMMA contents of **a** 27, **b** 60, and **c** 75 wt %

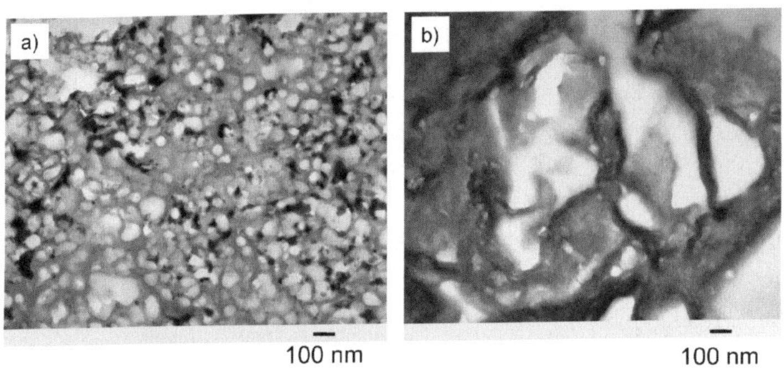

Fig. 13 TEM images of **a** homo-PE/homo-PMMA/PE-*b*-PMMA blended sample (homo-PE/homo-PMMA/PE-*b*-PMMA = 4/6/1 wt ratio) and **b** homo-PE/homo-PMMA blended sample (homo-PE/homo-PMMA = 4/6 wt ratio)

as chain transfer reagent, resulted in a binary catalyst system being oxidized using dry air at 100 °C, followed by hydrolysis with HCl to provide a terminally hydroxyl PE. It was converted to brominated PE using bromo-isobutyric bromide. PE-*b*-PnBA (M_n = 10 400, PE segment: 7.4 wt %) and Polyethylene-*block*-poly(*t*-butyl acrylate) (PE-*b*-PtBA) (M_n = 11 000, PE segment: 6.8 wt %) copolymers were obtained sucussessfully by ATRP.

PP-*b*-PMMA (M_n = 22 220, M_w/M_n = 1.14) was produced by CRP via another route. Terminally vinyl PP (M_n = 3100, M_w/M_n = 1.45, isotacticity = 32%) prepared using a zirconocene catalyst was converted to terminally brominated PP via PP-SiH prepared by hydrosilylation [70]. The resulting PP-*b*-PMMA was purified by extraction of unreacted PP with diethyl ether. Poly(ethylene-co-butene)-*block*-poly(methyl methacrylate) (EBR-*b*-PMMA) was synthesized through the bromination of terminally hydroxylated EBR (M_w = 3600 g/mol, M_w/M_n = 1.05), which was commercially available [71]. An atactic PP/PMMA had been synthesized by a combination of metallocene catalyses, Cp_2ZrCl_2 and $Me_2Si(CpMe_4)$(*N*-*t*-Bu)$TiCl_2$, and ATRP [72].

PP-*b*-PMMA and PP-*b*-PS block copolymers were produced using a brominated styrene-terminated PP, which was synthesized from styrene-terminated PP prepared with $Me_2Si(2\text{-}Me\text{-}4\text{-}NaphylInd)_2ZrCl_2$ [73].

For Graft Copolymers:

To synthesize graft copolymers by ATRP, the initiation site for living radical polymerization can be incorporated into the polymer chain. In general, copolymerization of olefins with functional monomers can be adopted for syntheses of PO macroinitiators.

PE graft copolymers were synthesized from PE-OH by Inoue et al. using ATRP techniques, adopting similar techniques as mentioned above [74]. PE-*g*-PMMA and Polyethylene-*graft*-poly(*n*-butyl acrylate) (PE-*g*-PnBA) were prepared through the combination of metallocene-catalyzed ethylene/10-undecen-1-ol copolymerization and conversion of the copolymer into PE-*g*-Br, as a macroinitiator, for ATRP. Well-defined graft copolymers, PE-*g*-PMMA and PE-*g*-PnBA, were confirmed by analyses of the detached side chains. Resulting PE-*g*-PMMA worked well as a compatibilizer.

The copolymer of ethylene with *p*-MS was available as a PE macroinitiator. PE-*g*-Br, which works as a PE macroinitiator for ATRP, was prepared from PE-*g*-MS by bromination with NBS [78], and PE-*g*-Br can be used to prepare PE-*g*-PMMA, PE-*g*-PS and PE-*g*-PnBA.

PP hybrids with grafted structures have been synthesized by Kaneko et al. adopting similar methods [76]. PP-g-OH prepared by metallocene ($Me_2Si(2\text{-}Me\text{-}4\text{-}PhInd)_2ZrCl_2$) -catalyzed copolymerization of propylene/10-undecen-1-ol was converted to PP-*g*-Br by bromination with 2-bromoisobutyryl bromide, which was used as a PP macroinitiator. The radical polymerization of methylmethacrylate, styrene, and *n*-butyl acrylate with the obtained PP macroinitiator using a CuBr/PMDETA catalyst system resulted in most of the consumed monomer being grafted onto the PP backbone following a soxhlet extraction test. The content of the polar segments in the obtained graft copolymers were controlled in the range of 11.6–46.6 wt %. For PP-*g*-PMMA and polypropylene-*graft*-polystyrene (PP-*g*-PS), the flexural modulus clearly increased in comparison with the base polymer (PP-*g*-OH), and the increase of the flexural modulus depended on the content of the grafted segment. On the other hand, for polypropylene-*graft*-poly(n-butyl acrylate) (PP-*g*-PnBA) the flexural modulus considerably decreased with the increase of PnBA content because of its softness. Figure 14 shows a plot of the flexural modulus and flexural strength versus the content of the polar segment. As in the case of the flexural modulus, the flexural strength has also been expected to increase by the grafting of a PMMA and a PS segment and to decrease by a PnBA segment. However, the flexural strength of PP-*g*-PS copolymers gradually decreased with the increase of PS content. This demonstrates that the balance of the mechanical properties for these new graft copolymers can

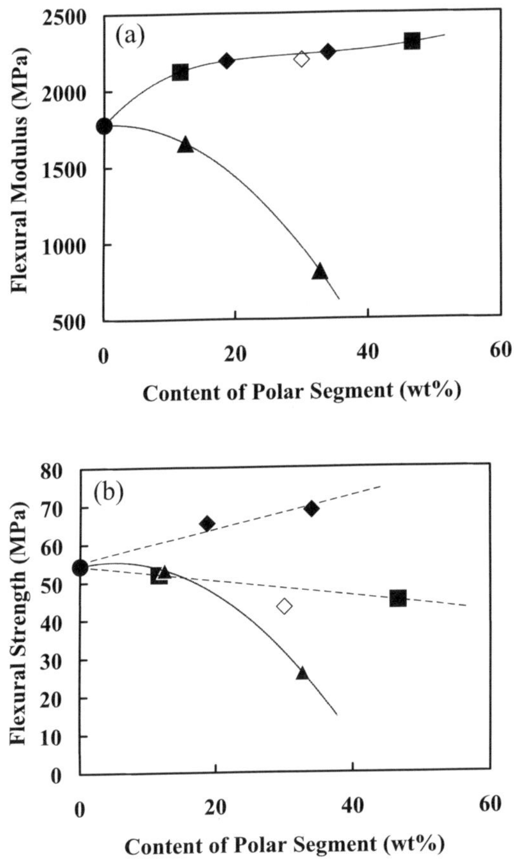

Fig. 14 Plot of **a** flexural modulus and **b** flexural strength versus content of polar segment (♦: PP-*g*-PMMA, ■: PP-*g*-PS, ▲: PP-*g*-PnBA, •: PP-*g*-OH, ◇: PP-*g*-OH/PMMA (7/3) Blend)

be controlled by the kind of grafting polar segment. Furthermore, since the flexural strength of the physical blended sample of PP-*g*-OH and PMMA (weight ratio = 7/3) was much lower than that of the obtained PP-*g*-PMMA graft copolymers, it has been confirmed that the chemical linkage between both segments significantly contributes to the enhancement of the mechanical properties. Finally, the result of Izod impact test is shown in Fig. 15. The Izod impact strength was remarkably improved by the incorporation of the PnBA segment, indicating the function of the PnBA segment as an impact modifier of PP. Thus, the polymer hybrids by chemical linkage between PP and other polymers are useful as a new material possessing unique and improved mechanical properties reflecting the kind and content of the grafted polar segments.

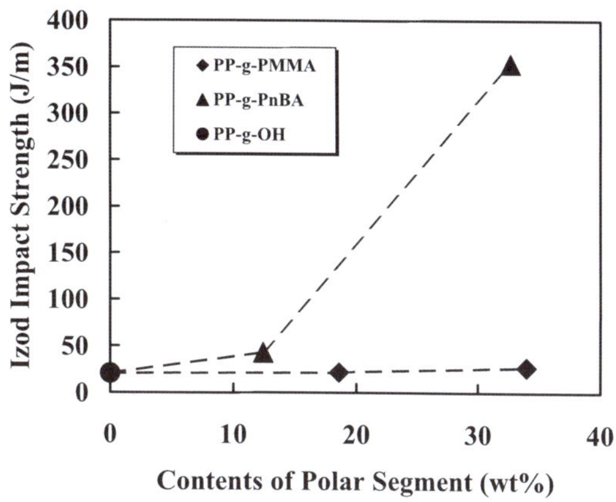

Fig. 15 Plot of Izod impact strength *versus* content of polar segment

The copolymerization of propylene with allyldimethylsilane (ADMS) was carried out with Ziegler-Natta catalyst systems supported on $MgCl_2$. The obtained PP-*g*-ADMS was treated with $SOCl_2$ to produce PP-*g*-Cl, which works as a macroinitiator for ATRP. PP-*g*-PMMA ($M_n = 67\,700$, $M_w/M_n = 5.3$, PMMA segment: 79.6 wt %) was produced from PP-Si-Cl with MMA and CuCl/PMDETA [77].

The synthesis of polyethylene-based graft copolymers by ATRP has been reported [78]. Poly(ethylene-*co*-styrene) prepared with a Ti complex catalyst system was used as starting material. The copolymer was converted to brominated poly(ethylene-*co*-styrene) as macroinitiator for ATRP by a bromination using NBS and AIBN. Resulting graft copolymers, PE/St-*g*-PS, PE/St-*g*-(PMMA-*b*-PS), PE/St-*g*-(PMMA-*b*-PMA), and PE/St-*g*-(PMMA-*b*-PHEMA), were characterized by NMR, DSC, GPC analyses, and scanning electron microscopy (SEM) images. PE-*g*-PS and PE-*g*-PMMA graft copolymers were prepared through the conversion of commercially available ethylene/(glycidyl methacrylate) copolymer to halogenated copolymers [79].

EPDM-*g*-PMMA was produced through ATRP with CuBr/bipyridine [80]. The graft copolymer, which had an ethylene-propylene terpolymer (EPDM) backbone and PMMA branches, was prepared from brominated EPDM that was produced with NBS to introduce allylbromine moiety on the backbone. Resulting EPDM-*g*-PMMA graft copolymers were characterized by solvent extraction, infrared (IR), and NMR techniques.

2.1.4
Reversible Addition-Fragmentation Chain Transfer Polymerization

Concerning the reversible addition-fragmentation chain transfer (RAFT) polymerization, which is a metal free CRP method (Fig. 16) [81–85], several

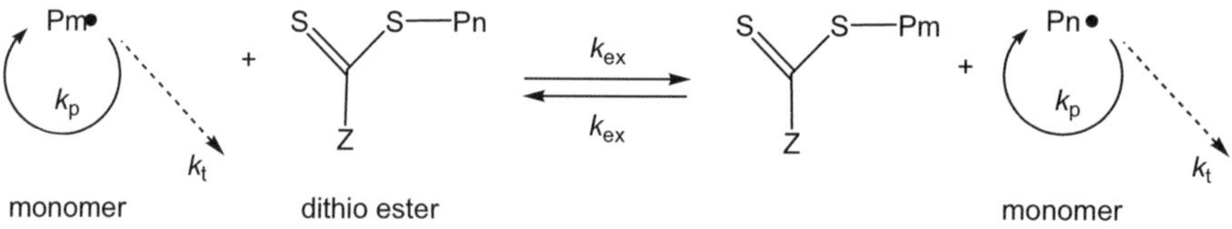

Fig. 16 RAFT polymerization mechanism

polymer hybrids based on polystyrene, cellulose, FEP film, and silica surface, have already been reported [86–91]. The growth of PS on PP via the RAFT process has been reported [92]. The γ-ray irradiation generates a radical on the surface of PP. Then, polymerization of styrene is initiated. Control of polymerization was performed by the presence of cumyl phenyldithioacetate (CPDA) as a chain transfer agent (CTA).

For Block Copolymers:

It has been proposed that the RAFT process was available to synthesize block copolymer containing PO segments. Polyethylene containing thio moiety would be obtained using Grignard chemistry. It could be shown that a pseudo-living ethylene polymerization via chain transfer reaction between dialkylmagnesium derivatives and a polymerization catalyst was carried out. Di-polyethylene magnesium compounds, obtained by a transition metal catalyzed chain growth reaction on butyloctyl magnesium, were reacted with a range of disulfides of thiocarbonilated compounds. These PEs with dithiocarbonate, dithiocarbamate, dithioester, and trithiocarbonate, which were macromonomer agents for RAFT polymerization, would serve as building blocks for the synthesis of block copolymers based on POs and polar segments. They were synthesized PE-*b*-PnBA from di-polyethylene magnesium compounds [93].

For Graft Copolymers:

The RAFT polymerization technique has been applied to synthesize polymer hybrids based on POs, according to Kawahara et al. [94]. A PE-PMMA polymer hybrid was synthesized using PE-CTA, which was prepared by sequential functionalization of terminally hydroxylated PE (Fig. 17). The structure of PE-CTA was confirmed by ^{1}H NMR and FT-IR analyses. The results of GPC after MMA polymerization revealed that the molecular weight (M_w) of the purified polymers increased from 36 600 by PE-CTA, to 38 500 and 40 000, respectively. ^{1}H NMR analysis of the purified polymers confirmed that the amounts of PMMA segments were in a range of 7.8 and 23 wt %. TEM micrographs indicated the nanometer level microphase-separation morphology between the PE segment and PMMA segment (Fig. 18), suggesting that poly-

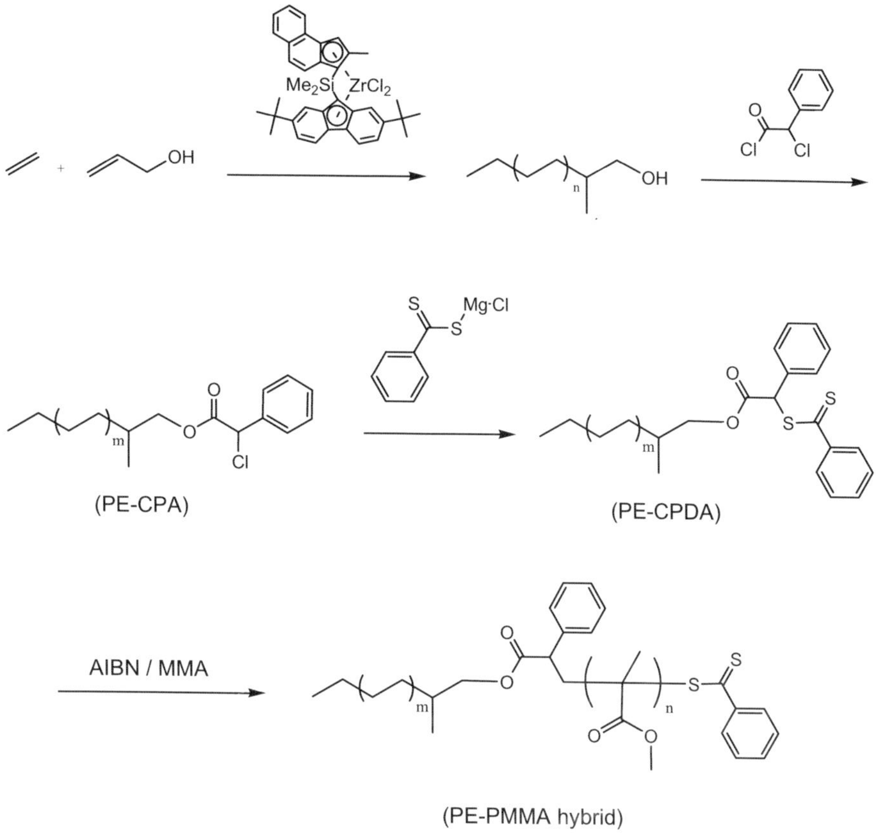

Fig. 17 Synthetic route of PE-PMMA polymer hybrid by RAFT polymerization method

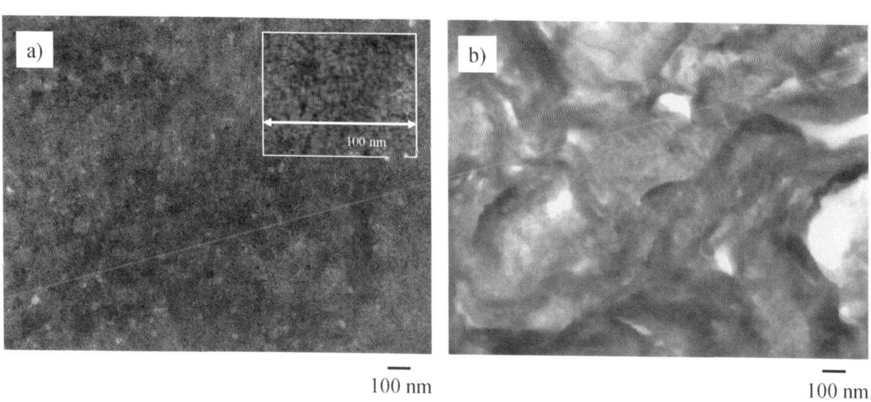

Fig. 18 TEM images of polymer sheets; **a** PE-PMMA purified polymer (PMMA contents; 23 wt %), **b** homo-PE/homo-PMMA blended polymer (homo-PE/homo-PMMA = 75/25 wt ratio)

mer hybrids based on PE were synthesized via the RAFT polymerization method.

2.1.5
Nitroxide Mediated Polymerization

Nitroxide mediated polymerization (NMP) is another type of controlled radical polymerization technique used to synthesize polymer hybrids. It relies on the reversible trapping of growing macro-radicals by nitroxide to form dormant species in which the C–ON covalent bond is thermally cleavaged (Fig. 19). At a polymerization temperature, the equilibrium between dormant and active species is strongly shifted to the dormant side, which limits the irreversible chain termination reaction.

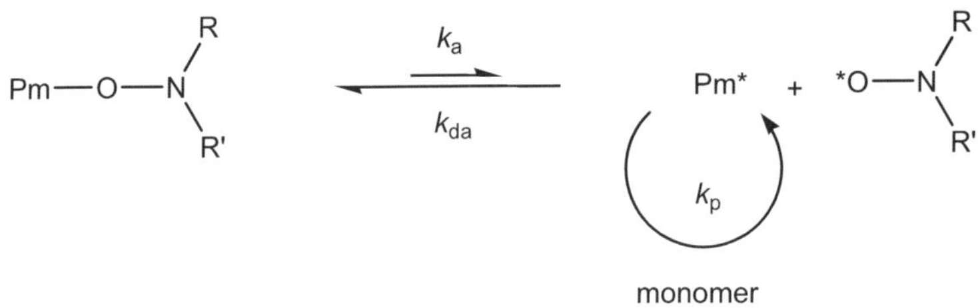

Fig. 19 NMP mechanism

The transformation of PP chains into macroalkoxyamine based on 2,2,6,6-tetramethyl-1-piperidinyl-1-oxy (TEMPO) was reported [95–98]. Commercial PP was irradiated by γ-rays to form peroxide groups into the polymer chain. Increasing the reaction temperature to 125 °C with a styrene monomer, copolymers based on PP and exhibiting PS or P(St-*co*-MMA) grafts were obtained.

For Block Copolymers:

The incorporation of TEMPO alkoxy amine at the end of a PE chain has been achieved [99]. The dialkylmagnesium compound in ethylene polymerization was adopted as a chain transfer agent, as mentioned above. It was also reported that PE-TEMPO and terminally *N*-(2-methyl-2-propyl)-*N*-(1-diethylphosphono-2,2-dimethylpropyl)-*N*-oxyl PE were synthesized by the reaction with di-polyethylene magnesium produced in ethylene polymerization. They were used for CRP of *n*-butyl acrylate, leading to PE-*b*-PnBA.

For Graft Copolymers:

A pure monomolecular initiation system would be more appropriate for the synthesis of graft copolymers. For the copolymerization of olefins with functional monomers, such as 1-(4-oxa-2-phenyl-11-dodeceneoxy)-2,2,6,6-tetramethylpiperidine or *N*-[4-(1-{[*tert*-butyl(2-methyl-1-phenylpropyl)amino]oxy}ethyl)benzyl]-*N*-cyclohexyl-*N*-pent-4-enylamine, metallocene catalysts have taken advantage of the potential [100]. PP-*g*-PS and poly(4-methylpent-1-ene)-*graft*-polystyrene (P4MP1)-*g*-PS) were synthesized from PP-TEMPO and P4MP1-TEMPO compounds, respectively. Also, poly(ethylene-*co*-butene)-*graft*-polystyrene (EBR-*g*-PS), poly(ethylene-*co*-butene)-*graft*-polybutadiene, and poly(ethylene-*co*-butene)-*graft*-poly(butyl acrylate) (EBR-*g*-PBA) were prepared. A highly branched PE graft copolymer, PE-*g*-PS, was synthesized from the copolymer of ethylene with 1-(4-oxa-2-phenyl-11-dodeceneoxy)-2,2,6,6-tetramethylpiperidine [101].

PE-*g*-PS and PP-*g*-PS were synthesized from PE-TEMPO and PP-TEMPO, respectively, which were prepared from PE and PP reacting with benzoyl peroxide and TEMPO [102].

Graft copolymers were also prepared by a combination of metallocene polymerization and living radical polymerization techniques [103]. Propylene was copolymerized with vinyl monomer bearing alkoxyamine using a $Et(THInd)_2ZrCl_2/B(C_6F_5)_4^-$ catalyst system. The alkoxyamine group present as a pendant group was utilized as a living radical initiator for styrene to give PP-*g*-PS with high molecular weight ($M_n = 210\,000$).

2.2
Polymerization with Macromonomers

In general, macromonomers possessing a polymerization chain end moiety were reported to be useful in producing graft copolymers with polymer branches. For instance, PO macromonomer possessing an acryl or metacryl chain end moiety would be useful for radical (controlled radical or free radical) and ionic polymerization. In this case, the polymer backbone was of a non-PO nature, and PO segments were graft chain. On the other hand, non-PO macromonomers possessing a vinyl chain end moiety would be useful for olefin polymerization. In that case, the polymer backbone was PO, and the non-PO segment would be graft chain and non-PO macromonomer used as a comonomer in the polymerization.

2.2.1
Polyolefin Macromonomer

A PP macromonomer with a methacryloyl end group was synthesized, and was used to prepare PMMA-*g*-PP graft copolymers by conventional free radical copolymerization [104]. Vinylidene-terminated PP ($M_n = 1000$) was converted into terminally hydroxylated PP (PP-OH) by a combination of the hydroboration reaction of the unsaturated group and oxidation reaction. Resulting PP-OH was reacted with methacryloylchloride to synthesize terminally methacrylated PP (PP-MA). PMMA-*g*-PP graft copolymers have been synthesized with MMA and 2,2′-azobis(isobutyronitrile) (AIBN).

The PnBA-*g*-PE graft copolymer was successfully prepared by a combination of two living polymerization techniques, Pd-mediated living ethylene polymerization and ATRP [105]. The living polymerization of ethylene with a Pd α-diimine catalyst afforded highly branched PE macromonomers end capped with methacrylate functional groups ($M_n = 10\,000$, $M_w/M_n = 1.1$). The PE macromonomer was then copolymerized with n-butyl acrylate by ATRP. The morphologies of the resulting graft copolymers were observed by atomic force microscopy (AFM).

Kaneko et al. have reported the preparation of EPR macromonomers, which is useful to synthesize PMMA-*g*-EPR graft copolymers [106]. Graft copolymers possessing a poly(methacrylate) backbone and EPR branches were successfully synthesized by the copolymerization of EPR macromonomers with MMA. Methacryloyl terminated EPR, which were EPR macromonomers, were prepared by the sequential end functionalization of EPRs with vinylidene end group via hydroalumination, oxidation, and esterification (Fig. 20). The ATRP method can be used for the copolymeriza-

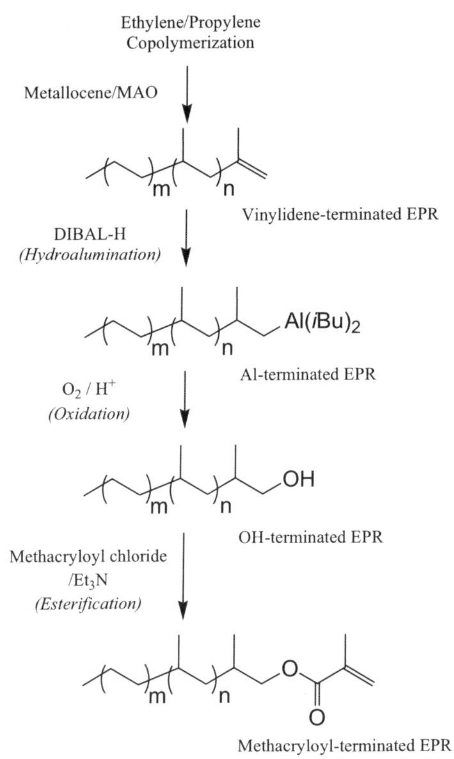

Fig. 20 Synthetic route of EPR macromonomer

tion reaction to give the PMMA-*g*-EPR graft copolymers with various EPR contents (8.6–38.1 wt %) and EPR branch numbers (1–14 branches) (Fig. 21, Table 2).

Furthermore, by using the ATRP method with a tetrafunctional initiator, the star copolymers consisting of four PMMA-*g*-EPR graft copolymer arms were also synthesized (Fig. 22). These unique topologies of the obtained copolymers can be confirmed by GPC measurement and NMR analysis. From TEM observation, the morphologies of these graft and star copolymers were remarkably altered by changing EPR branch number and the structure of the

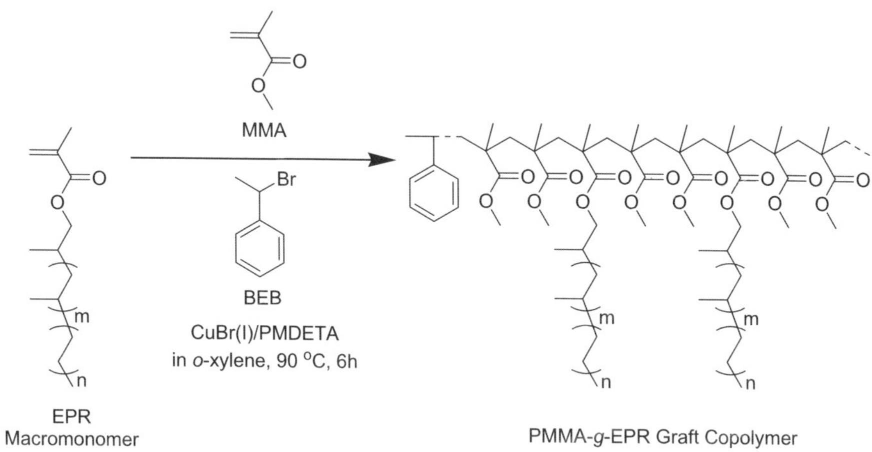

Fig. 21 Synthesis of PMMA-*g*-EPR graft copolymer

Table 2 Preparation of PMMA-*g*-EPR graft copolymers [a]

Copolymer	EPR macromonomer	MMA (ml)	[MMA]/[EPR]	Yield (g)	M_n [b]	M_w/M_n [b]	EPR content [c] (mol%)	(wt%)
G1	EPR M (0.9k)	4.28	10	2.77	28 200	1.40	6.2	38.1
G2	EPR M (0.9k)	4.28	25	2.52	27 500	1.41	2.6	19.9
G3	EPR M (0.9k)	2.14	50	1.06	29 300	1.39	2.0	15.9
G4	EPR M (0.9k)	2.14	100	1.01	32 300	1.45	1.0	8.6
G5	EPR M (3k)	2.14	50	1.49	27 700	1.48	1.1	24.9
G6	EPR M (3k)	2.14	80	1.19	34 600	1.56	0.72	18.0
G7	EPR M (3k)	2.14	200	1.00	29 000	1.62	0.37	10.0

[a] Polymn. Conditions: [BEB]/[CuBr]/[PMDETA]/[MMA] = 0.01/0.01/0.02/2 0 M in *o*-xylene, 90 °C × 6 h
[b] Determined by GPC at 40 °C in $CHCl_3$ and calibrated with PS
[c] Determined by ^1H NMR

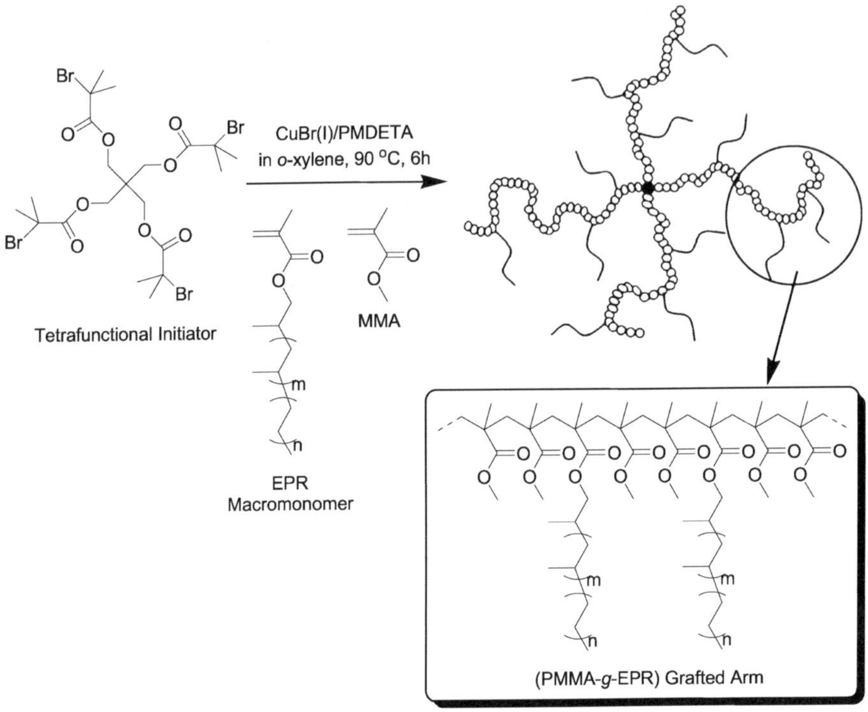

Fig. 22 Synthesis of the four arm (PMMA-*g*-EPR) star copolymer

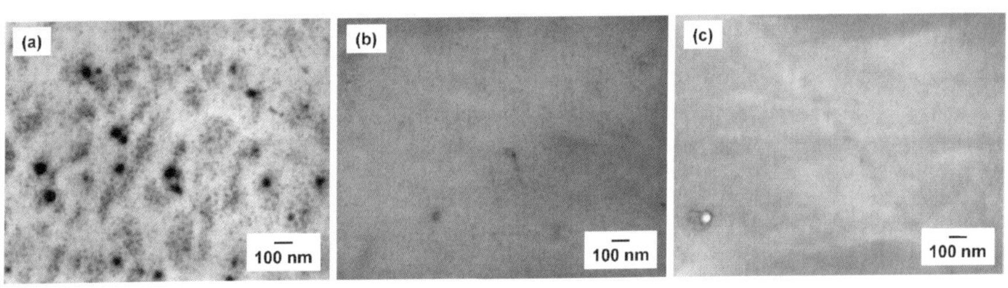

Fig. 23 TEM images of **a** PMMA-*g*-EPR graft copolymer (EPR content; 8.6 wt %), **b** PMMA-*g*-EPR graft copolymer (EPR content; 38.1 wt %), **c** four-arms (PMMA-*g*-EPR) star copolymer (EPR content; 36.3 wt %)

PMMA backbone, which can control the segmental dispersion of PMMA and EPR segments in the range from 1 nm to 100 nm (Fig. 23). Moreover, it was clearly demonstrated that the PMMA-*g*-EPR graft copolymer with 38.1 wt % of EPR content worked efficiently as a compatibilizer for an EPR/PMMA polymer blend (Fig. 24). DSC analysis revealed the effect of the EPR branch on the T_g value of the PMMA backbone and a little incorporation of an EPR branch caused a large deviation of the T_g value from the homo-PMMA.

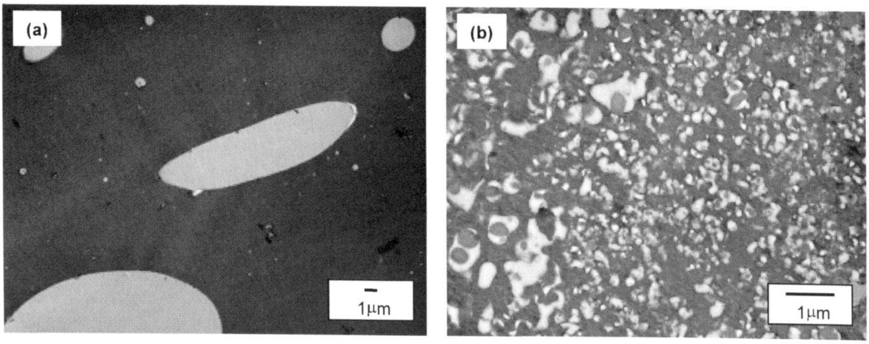

Fig. 24 TEM images of EPR/PMMA blend samples **a** before adding PMMA-*g*-EPR graft copolymer (EPR/PMMA = 5/5 wt ratio) and **b** after adding PMMA-*g*-EPR graft copolymer (EPR/PMMA-*g*-EPR/PMMA = 45/10/45 wt ratio)

On the other hand, the PMMA-*g*-EPR graft copolymers using the FRP method with AIBN showed a different morphology and thermal property from those after using the ATRP method, leading to the fact that the ATRP method is preferable compared to FRP method for the purpose of preparing well-defined macromolecular structures. Poly(EPR) can also be prepared by the free radical polymerization with AIBN [107].

The preparation of PnBA-*g*-PE and PtBA-*g*-PE graft copolymers was reported using Fe-mediated olefin polymerization, chain shuttling with Zn and ATRP techniques [108]. Terminally hydroxyl PE was synthesized from Zn-terminated PE by oxidation and hydrolysis, as referred to above. It was converted to methacrylated PE, as PE macromonomer, using methacryloyl chloride. The resulting PE macromonomer was used for the copolymerization of nBA or tBA by ATRP using CuBr/tris((*N*,*N*-dimethylamino)ethyl)amine. The obtained graft copolymers were characterized by GPC, DSC, and ^{1}H NMR.

2.2.2
Polystyrene Macromonomer

PS macromonomers have been efficiently applied to the synthesis of well-defined polymer hybrids with controlled length of grafts. They are, in general, prepared via living anionic polymerization of styrene monomers and their treatment with vinyl compounds, such as ω-allyl, ω-undecenyl, and styryl compounds.

Far better results have been obtained for the copolymerization of propylene and ω-allyl PS macromonomers, prepared by *s*-BuLi, using a zirconocene catalyst, $Me_2Si(2\text{-}Me\text{-}4,5\text{-}BzInd)_2ZrCl_2$/MAO. At high metallocene catalyst concentrations, the activity is rather high, but only 7 wt % of the PS macromonomer was incorporated. At lower catalyst concentrations, the incorporation of PS macromonomer increased up to 72 wt % with a decrease of the catalytic activity [109].

ω-Allyl PS macromonomers, which were synthesized by the ATRP of styrene with CuBr/bipyridine, have been used as comonomers in metallocene-catalyzed propylene copolymerizataions using $Me_2Si(2\text{-}Me\text{-}4,5\text{-}BzInd)_2ZrCl_2$/MAO [110]. It has been found that the incorporation of the PS macromonomers increases with a decrease in molar mass of the macromonomer and propylene concentration and increasing polymerization temperature. The highest comonomer incorporation (10.8 wt %) was achieved in the copolymerization at 70 °C.

The synthetic pathway of PE-*g*-PS graft copolymer via the coordination copolymerization of ethylene with PS macromonomers has been prepared using a Pd complex [111]. ω-Vinylbenzyl PS macromonomers could not be incorporated into PE chains. On the other hand, the incorporation of ω-allyl PS macromonomers was achieved. It has also been reported on the comparison of the ease of incorporation of PS macromonomers among ω-allyl, ω-undecenyl, and α,ω-undecenyl PS macromonomers into PE chains [112]. It was shown that the undecenyl end group was more reactive than an allyl end group.

PP graft copolymers were synthesized by a PS macromonomer method. Norbornenyl terminated PS macromonomers prepared by anionic polymerization of styrene (M_w = 800–18 500) were copolymerized with propylene using a $Me_2Si(2\text{-}Me\text{-}4,5\text{-}BzInd)_2ZrCl_2$/MAO catalyst to give PP-*g*-PS [113]. The maximum incorporation of PS macromonomer was 69 wt %. Resulting PP-*g*-PS showed a good compatibilizing effect in PP and PS blends.

2.3
Coupling Reaction with Reactive Polyolefin

Reactive POs having reactive functional groups are considered to be the most conventional way to produce various polymer hybrids, most of which were graft type copolymers. Typically, it is well-known that the maleic anhydride modified POs are useful as a reactive PO to prepare the polymer hybrid [114].

A synthetic route to graft polyolefinic block copolymers have been investigated by Kashiwa et al. with a polymer coupling reaction as an application of our expertise in synthesizing terminally hydroxylated PP (PP–OH) [115]. In previous papers, a hydroxyl group of PP–OH is expected to show high reactivity to maleic anhydride modified polymers [116–118]. The PP-OH was used for the coupling reaction with maleic anhydride modified PE (PE-*g*-MAH) to synthesize polyolefin hybrid, PE-*g*-PP [119]. For its comparison, PP was blended with PE-*g*-MAH under the same conditions as the coupling reaction. As shown in Fig. 25, a peak in the low molecular weight region of the coupling product is relatively reduced compared to that of the polymer blend and, in its place, the coupling product has the larger peak in the high molecular weight

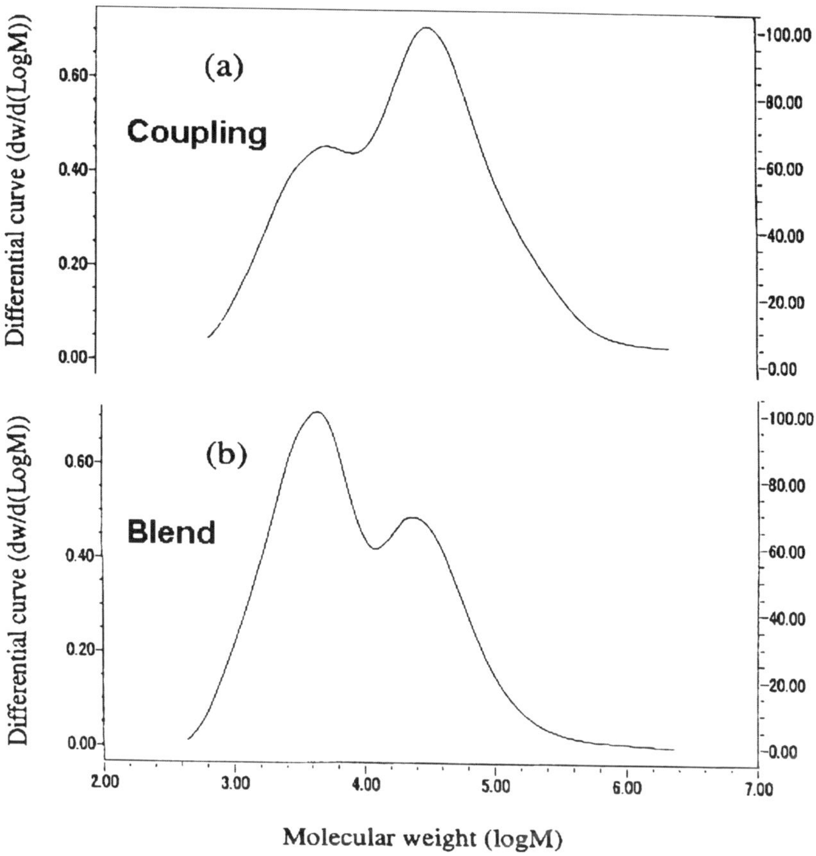

Fig. 25 GPC diagrams of **a** a product by coupling reaction between PP-OH and PE-*g*-MAH and **b** a polymer blend consisting of py-PP and PE-*g*-MAH

region. It would be indicated that PE-*g*-PP was produced by coupling reaction between –OH and -MAH units. They also reported the coupling reaction between EBR-*g*-MAH and PP–OH to prepare polyolefin hybrid, EBR-*g*-PP.

It has been reported on the synthesis method of PP-*g*-PS by a coupling reaction by Kawahara et al. At first, they synthesized propylene/11-bromoundec-1-ene copolymer (PP-Br) using an En(Ind)$_2$ZrCl$_2$/MAO catalyst system. Then, PP-*g*-PS was synthesized by the coupling reaction between PP-Br and polystyryl lithium salts (PS-Li) prepared by living anionic polymerization of styrene with *n*-butyl lithium (*n*-BuLi) (Fig. 26) [120]. Structures of the resulting polyolefin hybrids were confirmed by ^{1}H NMR analysis. TEM micrographs of PP-*g*-PS copolymers indicated the nanometer level microphase-separation morphology between the polypropylene segment and the polystyrene segment. Obtained PP-*g*-PS copolymers is expected to work as an effective compatibilizer for PP and PS. To estimate this effectiveness,

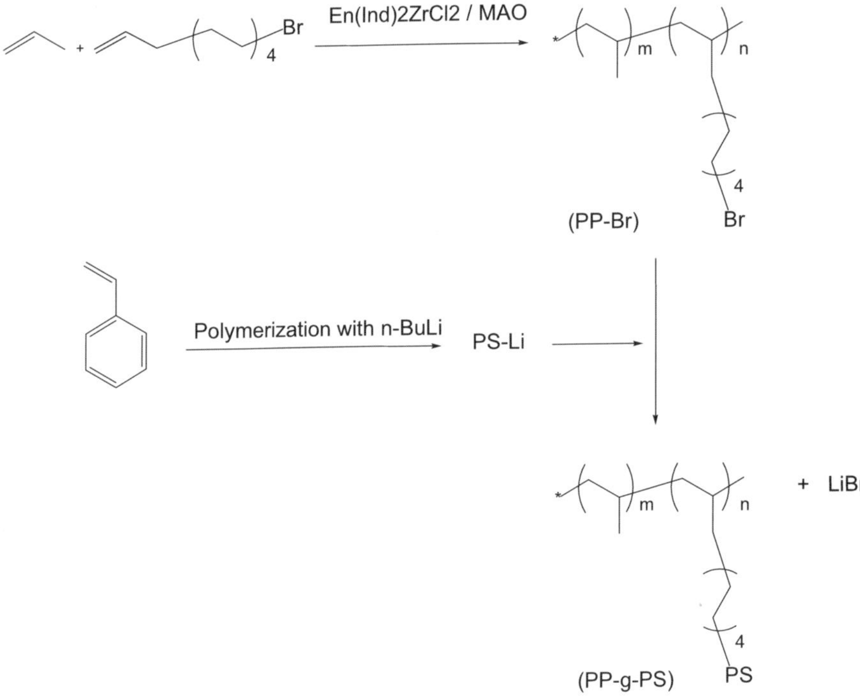

Fig. 26 Synthetic route of PP-*g*-PS copolymers by coupling reaction

the PP-*g*-PS copolymer and homo-PP and homo-PS (homo-PP/homo-PS/PP-*g*-PS = 5/5/1 wt ratio) were blended at 130 °C in *o*-xylene. For comparison, a blended sample without PP-*g*-PS (homo-PP/homo-PS = 5/5 wt ratio) was also prepared. Figure 27 shows TEM images of these blended polymers. The blended polymers containing PP-*g*-PS showed morphology in which the size of the dispersed domains was smaller than that of the blended polymer without PP-*g*-PS. From the TEM images, it can be suggested that the size of the domain in the PP/PS blended polymer depended on the PS contents of the PP-

Fig. 27 TEM images of PP/PS (5/5 wt ratio) polymer sheets **a** without PP-*g*-PS, **b** with PP-*g*-PS (PS content; 20 wt %), **c** with PP-*g*-PS (PS content; 51 wt %)

g-PS. These results indicate that such PP-*g*-PS copolymers effectively improve the compatibility between PP and PS.

PE-*g*-PS graft copolymer was produced by a coupling reaction, too. α-Carboxyl PS prepared by an ATRP technique was reacted with PE-*g*-glycidyl methacrylate (GMA) to produce PE-*g*-PS [121]. A PP-*b*-PMMA block copolymer was synthesized using a magnesium bromide terminated PP as an initiator for the radical polymerization of MMA, which was prepared from the vinylidene terminated PP obtained with the $Et(Ind)_2ZrCl_2$/MAO catalyst system [122].

2.4
Living Polymerization of Olefins

Using living conditions with the state-of-the-art in catalyst technologies, crystal-amorphous polyolefin hybrids can be synthesized by successive polymerization, for instance, homo-polymerization and co-polymerization, of olefin monomers. By this methodology, block copolymers of polymer hybrids can be obtained.

Bis(phenoxy-imine)metal complexes opened the way to the synthesis of olefin block copolymers. Polyolefin hybrids, such as polyethylene-*block*-*syndiotactic*polypropylene (PE-*b*-*s*PP), polyethylene-*block*-poly(ethylene-co-propylene) (PE-*b*-EPR), *s*PP-*b*-EPR, PE-*b*-EPR-*b*-*s*PP, PE-*b*-EPR-*b*-PE, and polyethylene-*block*-poly(ethylene-co-hexene) have been synthesized with those complexes via living olefin (co)polymerizations [123–128]. Morphology of PE-*b*-EPR was also studied by atomic force microscopy (AFM) images [129]. The morphology and thermodynamic behavior of *s*PP-*b*-EPR produced with bis(salicylaldiminate)titanium complex/MAO system was studied using DMS, DSC, and TEM as analytical methods [130, 131].

The synthesis of PP-*b*-EPR can be accomplished by a stopped-flow polymerization method, whose polymerization time is very short and which is a quasi-living system, in the presence of a $MgCl_2$-supported titanium catalyst [132–135]. The results of GPC, ^{13}C NMR, CFC, DSC, and optical microscopic observation indicated the formation of a block copolymer having a chemical linkage between PP and EPR segments.

Polyethylene-*block*-poly(ethylene-*co*-norbornene) (PE-*b*-P(E-*co*-NBE)) block copolymer was successfully synthesized by a titanium complex with two non-symmetric bidentate β-enaminoketonato ligands [136, 137]. Bis(pyrrolide-imine)titanium complex also has the ability to produce the PE-*b*-P(E-*co*-NBE) block copolymer. PE-*b*-PS was synthesized via sequential monomer addition during homogeneous polymerization with bis(phenoxy-imine)metal catalysts [138].

A zirconium-based post-metallocene catalyst can produce PP-*b*-PE. This has been reported as ^{13}C NMR evidence of the real block nature for a polyolefin based materials [139].

Controlled block copolymerization of olefins with polar monomers was performed with a lanthanide complex by the successive polymerization of hexene (or pentene) and methylmethacrylate (or caprolactone). Polyhexene-*block*-poly(methyl methacrylate), polyhexene-*block*-polycaprolactone, poly-pentene-*block*-poly(methyl methacrylate), and polypentene-*block*-polycaprolactone were synthesized using a lanthanide complex as initiator [140–143].

A Ni α-diimine catalyst was adopted to synthesize elastmeric block copolymers, such as polypropylene-*block*-polyhexene and polyoctadece-*block*-poly(propylene-*co*-octadecene)-*block*-polyoctadecene [144].

It has also been demonstrated that a PP-*b*-EPR block copolymer was prepared with a living polymerization catalyst composed of $V(acac)_3/AlEt_2Cl$ at −78 °C [145]. Also, PP-*b*-EPR was produced with metallocene catalyst systems at −78 °C [146].

The production of olefin block copolymers with alternating semicrystalline and amorphous segments has been achieved by varying the ratio of α-olefin to ethylene in the two types of block segments [147]. The catalyst system used a chain shuttling agent to transfer growing chains between two distinct catalysts with different monomer selectivities in a single polymerization reactor. The block copolymers, which were probably multi-block copolymers, simultaneously have high melting temperatures and low glass-transition temperatures, and, therefore, they maintained excellent elastic properties at high temperatures. Furthermore, the materials were effectively produced in an economically favorable, continuous polymerization process.

3 Polyolefin Hybrids Applications

As mentioned, polymer hybrids based on POs are effective as a compatibilizer between the olefinic materials and polar ones. Furthermore, some polymer hybrids, such as PP-*g*-PMMA, etc., show good mechanical strength as polymer materials. On the other hand, surface modification of the molded polymer is one of the most attractive methods to let polyolefin materials functionalize. In this sense, surface polymerization of functional monomers on polyolefins is an important subject for polyolefin hybrids. As previously referred to, the growth of PS on PP via the RAFT process has been reported [92].

Surface polymerization of 2-hydroxyethyl methacrylate (HEMA), which possessed a functional group, at the initiation sites on the molded sheet of the PP macroinitiator was performed by Matsuo et al. using the ATRP-mediated CuBr/PMDETA catalyst system [148]. The molded PP sheet ($d = 1.0$ mm) was prepared by a molding press from PP–Br, which was synthesized by bromination of propylene/10-undecen-1-ol copolymer (PP–OH; $M_n = 15\,500$, [–OH] = 3.8 units/chain). The ATRP of HEMA was carried out on the molded PP–Br

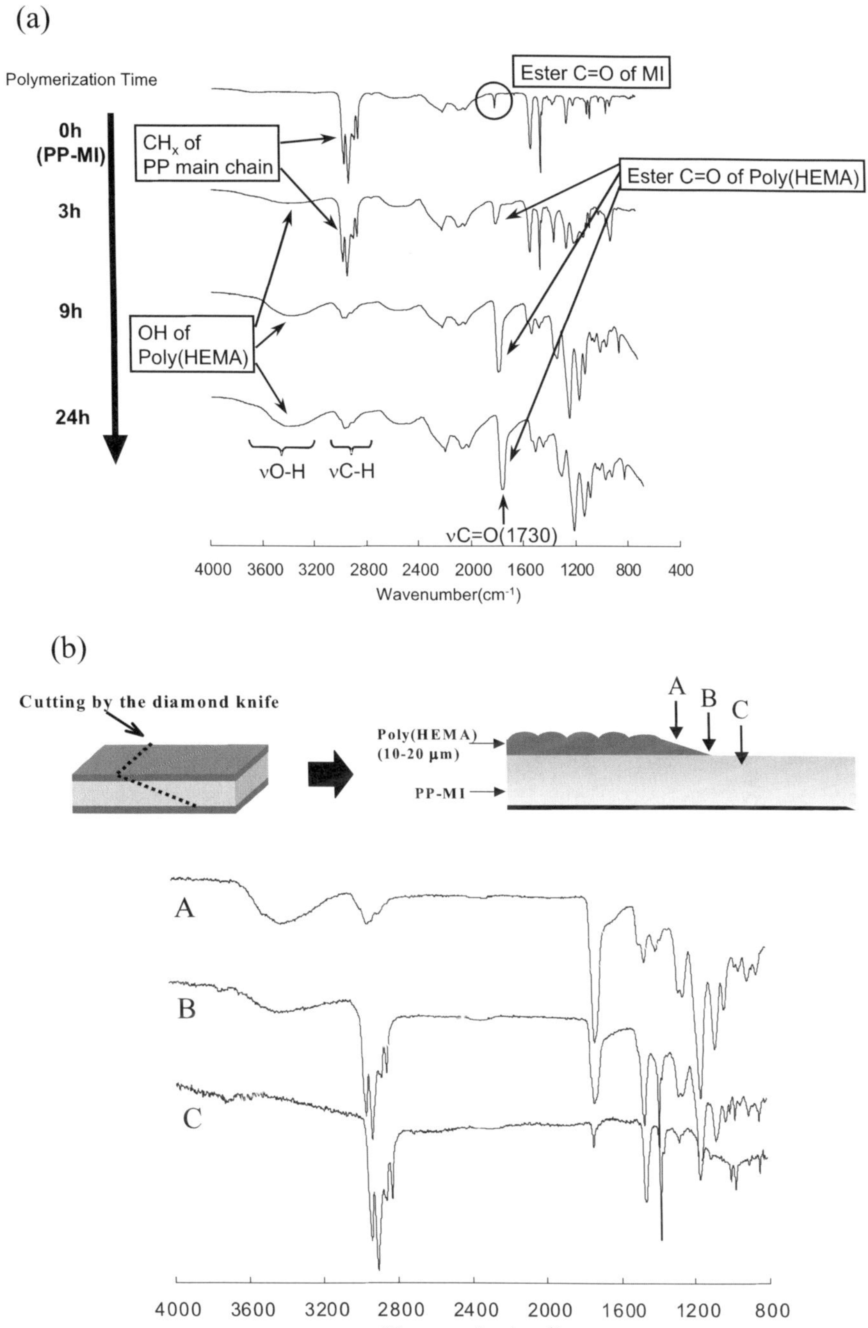

Fig. 28 ATR/IR spectra **a** on the surface region of the sheet, and **b** of the cross-section of PP-*graft*-Poly(HEMA) sheet

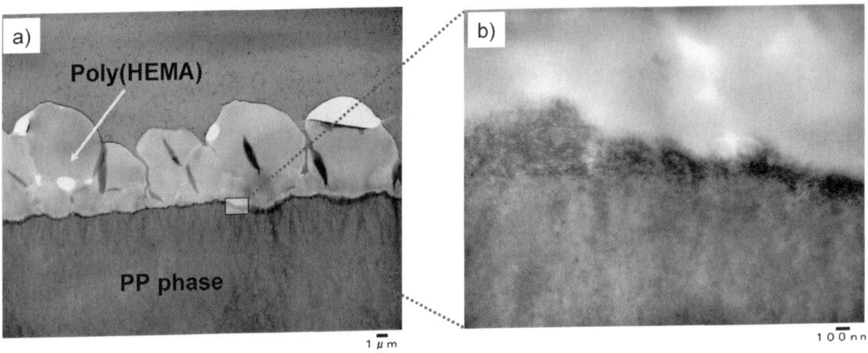

Fig. 29 TEM images of the section nearby the surface of PP-*graft*-Poly(HEMA) sheet at a magnification of **a** ×5000 and **b** ×60 000

sheet in ethanol with a CuBr/PMDETA catalyst system at ambient temperature.

The resulting sheets can be coated with poly(HEMA), then, analyzed by attenuated total reflection infrared (ATR/IR) (Fig. 28) and transmission electron microscopy (TEM) (Fig. 29) to investigate the structure and the morphology. During the polymerization of HEMA, according to ATR/IR analysis, it was observed that the peaks of the surface region were changed from the original PP–Br to that of PolyHEMA. Obviously, the features of the surface of the PP sheets were also changed after time. It was revealed that PP-*g*-Poly(HEMA) was successfully synthesized and showed unique morphological features.

This technique could be applied to introduce polar polymer segments onto the surface of the various molded parts based on POs, such as films, sheets, etc.

4 Conclusion

Here, we have reviewed recent advances in polymer-hybrid-based polyolefins regarding synthetic methodologies, structures, and their applications. Several methodologies, such as radical, anionic, cationic polymerizations, and post-polymerization reactions for the preparation of polyolefin hybrids have been reported. They are classified as PO macroinitiator, PO macromonomer, and reactive PO, except for the living copolymerization of olefins. These three tools offer a comprehensive coverage of polyolefin block or graft copolymers with well-defined molecular structures. Polyolefin hybrid materials possess novel and unique properties, due to the synergic effect of each segment having different characteristics. Actually, synthesized polymer hybrids based on POs possess excellent properties, such as compatibility and mechanical strength, compared to conventional POs. Furthermore, polymer hybrids based on POs seem to possess several advantages, namely, energy-saving in fabrication of polymers, process simplification, and saving of the oil resources by broadening the application of olefinic materials that are easily recyclable. The production of new polymer materials is also something of a revolution.

References

1. Kashiwa N (2003) J Polym Sci A Polym Chem 42:1
2. Yanjarappa MJ, Sivaram S (2002) Prog Polym Sci 27:1347
3. Bruzaud S, Cramail H, Duvignac L, Deffieux A (1997) Macromol Chem Phys 198:291
4. Stehling UM, Stein KM, Fischer D, Waymouth RM (1999) Macromolecules 32:14
5. Tsuchida A, Bolln C, Sernetz FG, Frey H, Mulhaupt R (1997) Macromolecules 30:2818
6. Zheng L, Farris J, Coughlin EB (2001) Macromolecules 34:8034
7. Wilen CE, Luttikhedde H, Hjertberg T, Nasman JH (1996) Macromolecules 29:8569
8. Wilen CE, Nasman JH (1994) Macromolecules 27:4051
9. Wilen CE, Auer M, Stranden J, Nasman JH, Rotzinger B, Steinmann A, King RE III, Zweifel H, Drewes R (2000) Macromolecules 33:5011
10. Mustnen I, Hukka T, Pakkanen T (2000) Macromol Rapid Comm 21:1286
11. Kaya A, Jakisch L, Komber H, Ponpe G, Pionteck J, Voit B, Schulze U (2000) Macromol Rapid Comm 21:1267
12. Kaya A, Jakisch L, Komber H, Voigt D, Pionteck J, Voit B, Schulze U (2001) Macromol Rapid Comm 22:972
13. Yang Y, Zhang F, Endo T, Hirotsu T (2003) Macromolecules 36:4709
14. Gayload NG, Mehta M (1982) J Polym Sci Polym Chem Ed 20:481
15. Rasumussen JR, Stedronsky ER, Whitesides GM (1977) J Am Chem Soc 99:4736
16. Bergbreiter DE, Franchina JG, Kabza K (1999) Macromolecules 32:4993
17. Poncin-Epaillard F, Brosse JC, Falher T (1997) Macromolecules 30:4415
18. Inagaki N, Tasaka S, Imai M (1993) J Appl Polym Sci 48:1963
19. Mülhaupt R, Duschek T, Rieger B (1991) Macromol Chem Macromol Symp 48/49:317
20. Shiono T, Soga K (1992) Macromol Chem Rapid Commun 13:371
21. Tsubaki S, Jin J, Sano T, Uozumi T, Soga K (2001) Macromol Chem Phys 202:1757
22. Shiono T, Kurosawa H, Ishida O, Soga K (1993) Macromolecules 26:2085
23. Shiono T, Kurosawa H, Soga K (1994) Macromolecules 27:2635
24. Worner C, Roseh J, Hohn A, Mülhaupt R (1996) Polym Bull 36:309
25. Shiono T, Azad SM, Ikeda T (1999) Macromolecules 32:5723
26. Markel EJ, Weng W, Peacock AJ, Deckmezian AH (2000) Macromolecules 33:8541
27. Chung TC (2002) Prog Polym Sci 27:39
28. Lopez RG, D'agosto F, Boisson C (2007) Prog Polym Sci 32:419
29. Doi Y, Watanabe Y, Ueki S, Soga K (1983) Macromol Chem Rapid Commun 4:533
30. Chung TC, Dong JY (2001) J Am Chem Soc 123:4871
31. Dong JY, Chung TC (2002) Macromolecules 35:1622
32. Chung TC, Xu G, Lu Y, Hu Y (2001) Macromolecules 34:8040
33. Chung TC, Lu B (1998) Macromolecules 31:5943
34. Lu B, Chung TC (1999) Macromolecules 32:2525

35. Chung TC, Lu HL, Janvikul W (1997) Polymer 38:1495
36. Chung TC (1988) Macromolecules 21:865
37. Lu Y, Hu Y, Wang ZM, Manias E, Chung TC (2002) J Polym Sci A Polym Chem 40:3416
38. Lu Y, Hu Y, Chung TC (2005) Polymer 46:10585
39. Han CJ, Lee MS, Byun DJ, Kim SY (2002) Macromolecules 35:8923
40. Chung TC, Lu HL (1997) J Polym Sci A Polym Chem 35:575
41. Chung TC, Lu HL (1998) J Polym Sci A Polym Chem 36:1017
42. Chung TC, Lu HL, Ding RD (1997) Macromolecules 30:1272
43. Lu HL, Chung TC (1999) J Polym Sci A Polym Chem 37:4176
44. Caporaso L, Iudici N, Oliva L (2006) Macromol Symp 234:42
45. Dong JY, Hong H, Chung TC, Wang HC, Datta S (2003) Macromolecules 36:6000
46. Chung TC, Dong JY (2002) Macromolecules 35:2868
47. Imuta J, Kashiwa N, Toda Y (2002) J Am Chem Soc 124:1176
48. Imuta J, Toda Y, Kashiwa N (2001) Chem Lett 710
49. Hagihara H, Tsuchihara K, Sugiyama J, Takeuchi K, Shiono T (2004) Macromolecules 37:5145
50. Kashiwa N, Matsugi T, Kojoh S, Kaneko H, Kawahara N, Matsuo S, Nobori T, Imuta J (2003) J Polym Sci A Polym Chem 41:3657
51. Chung TC, Rhubright D (1994) Macromolecules 27:1313
52. Chung TC, Janvikul W, Bernard R, Jiang GJ (1994) Macromolecules 27:26
53. Chung TC, Jiang GJ (1992) Macromolecules 25:4816
54. Chung TC, Janvikul W, Lu HL (1996) J Am Chem Soc 118:705
55. Xu G, Chung TC (1999) J Am Chem Soc 121:6763
56. Xu G, Chung TC (1999) Macromolecules 32:8689
57. Chung TC, Lu HL (1997) J Mol Cat A Chem 115:115
58. Ramakrishnan S, Berluche E, Chung TC (1990) Macromolecules 23:378
59. Chung TC, Rhubright D (1991) Macromolecules 24:970
60. Chung TC, Rhubright D (1993) Macromolecules 26:3019
61. Chung TC, Lu L, Li CL (1995) Polym Int 37:197
62. Chung TC, Janvikul W (1999) J Organomet Chem 581:176
63. Chung TC, Jiang GJ, Rhubright D (1994) US Patent 5 286 800
64. Chung TC, Jiang GJ, Rhubright D (1995) US Patent 5 401 805
65. Chung TC, Jiang GJ (1992) Macromolecules 25:4816
66. Chung TC, Rhubright D, Jiang GJ (1993) Macromolecules 26:3467
67. Matsugi T, Kojoh S, Kawahara N, Matsuo S, Kaneko H, Kashiwa N (2003) J Polym Sci A Polym Chem 41:3965
68. Inoue Y, Matyjazewski K (2004) J Polym Sci A Polym Chem 42:496
69. Kaneyoshi H, Inoue Y, Matyjazewski K (2005) Macromolecules 38:5425
70. Matyjazewski K, Saget J, Pyun J, Schlogl M, Rieger B (2002) J Macromol Sci A Pure and Appl Chem 39:901
71. Schellekens MAJ, Klumperman B, Linde R (2001) Macromol Chem Phys 202:1595
72. Dix A, Ptacek S, Poser S, Arnold M (2006) Macromol Symp 236:186
73. Zhang LY, Fan GQ, Guo CY, Dong JY, Hu YL, Huang MB (2006) Eur Polym J 42:1043
74. Inoue Y, Matsugi T, Kashiwa N, Matyjazewski K (2004) Macromolecules 37:3651
75. Hwu JM, Chang MJ, Lin JC, Cheng HY, Jiang GJ (2005) J Organomet Chem 690:6300
76. Kaneko H, Matsuo S, Kawahara N, Saito J, Matsugi T, Kashiwa N (2008) Macromol Symp 260:9
77. Li H, Zhao H, Zhang X, Lu Y, Hu Y (2007) Eur Polym J 43:109
78. Liu S, Sen A (2001) Macromolecules 34:1529

79. Matyjaszewski K, Teodorescu M, Miller PJ, Peterson ML (2000) J Polym Sci A Polym Chem 38:2440
80. Wang XS, Luo N, Ying SK (1999) Polymer 40:4515
81. Caporaso L, Iudici N, Oliva L (2005) Macromolecules 38:4894
82. Favier A, Charreyre MT (2006) Macromol Rapid Commun 27:653
83. Chiefari J, Chong YK, Ercolo F, Krstina J, Jeffery J, Le TPT, Mayadunne RTA, Meijs GF, Moad CL, Moad G, Rizzardo E, Thang SH (1998) Macromolecules 31:5559
84. Tujii Y, Ejaz M, Sato K, Goto A, Fukuda T (2001) Macromolecules 34:8872
85. Thomas DB, Convertine AJ, Myrick JL, Scales CW, Smith AE, Lowe AB, Vasilieva YA, Ayres N, McCormick CL (2004) Macromolecules 37:8941
86. Perrier S, Takolpuckdee P, Mars CA (2005) Macromolecules 38:6770
87. Roy D, Guthrie JT, Perrier S (2005) Macromolecules 38:10363
88. Stehling UM, Malmstrom EE, Waymouth RM, Hawker CJ (1998) Macromolecules 31:4396
89. Bowden NB, Dankova M, Wiyatno W, Hawker CJ, Waymouth RM (2002) Macromolecules 35:9246
90. Yoshikawa C, Goto A, Tsujii Y, Fukuda T, Yamamoto K, Kishida A (2005) Macromolecules 38:4604
91. Li C, Benicewicz BC (2005) Macromolecules 38:5929
92. Barner L, Zwaneveld N, Perera S, Pham Y, Davis TP (2002) J Polym Sci A Polym Chem 40:4180
93. Lopez RG, Boisson C, D'Agosto F, Spitz R, Boisson F, Gigmes D, Bertin D (2006) Macromol Rapid Comm 27:173
94. Kawahara N, Kojoh S, Matsuo S, Kaneko H, Matsugi T, Kashiwa N (2006) Polym Bull 57:805
95. Miwa Y, Yamamoto K, Sakaguchi M, Shimada S (1999) Macromolecules 32:8234
96. Miwa Y, Yamamoto K, Sakaguchi M, Shimada S (2001) Macromolecules 34:2089
97. Yamamoto K, Nakazono M, Miwa Y, Hara S, Sakaguchi M, Shimada S (2001) Polym J 39:1120
98. Yamamoto K, Tanaka H, Sakaguchi M, Shimada S (2003) Polymer 44:7661
99. Lopez RG, Boisson C, D'agosto F, Spitz R, Boisson F, Gigmes D, Bertin D (2007) J Polym Sci A Polym Chem 45:2705
100. Bowden NB, Dankova M, Wiyatno W, Hawker CJ, Waymouth RM (2002) Macromolecules 35:9246
101. Baumert M, Heinemann J, Thomann R, Mülhaupt R (2000) Macromol Rapid Commun 21:271
102. Park ES, Jin HJ, Lee IM, Kim MN, Lee HS, Yoon JS (2002) J Appl Polym Sci 83:1103
103. Lopez RG, Boisson C, D'agosto F, Spitz R, Boisson F, Bertin D, Tordo P (2004) Macromolecules 37:3540
104. Duschek T, Mülhaupt R (1992) Polym Prepr 33:170
105. Hong SC, Jia S, Teodorescu M, Kowalewski T, Matyjaszewski K, Gottfried AC, Brookhart M (2002) J Polym Sci A Polym Chem 40:2736
106. Kaneko H, Kojoh S, Kawahara N, Matsuo S, Matsugi T, Kashiwa N (2005) J Polym Sci A Polym Chem 43:5103
107. Kaneko H, Kojoh S, Kawahara N, Matsuo S, Matsugi T, Kashiwa N (2004) Macromol Symp 213:335
108. Kaneyoshi H, Inoue Y, Matyjaszewski K (2005) Macromolecules 38:5425
109. Henschke O, Neubauer A, Arnold M (1997) Macromolecules 30:8097
110. Schulze U, Fonagy T, Komber H, Pompe G, Pionteck J, Ivan B (2003) Macromolecules 36:4719

111. Lahitte JF, Peruch F, Plentz-Meneghetti S, Isel F, Lutz PJ (2002) Macromol Chem Phys 203:2583
112. Lahitte JF, Plentz-Meneghetti S, Peruch F, Isel F, Muller R, Lutz PJ (2006) Polymer 47:1063
113. Arnold M, Knorr J, Köller F, Bornemann S (1999) J Macromol Sci A Pure and Appl Chem A36:1655
114. Moad G (1999) Prog Polym Sci 24:81
115. Johnson LK, Killian CM, Brookhart MS (1995) J Am Chem Soc 117:6414
116. Younkin TR, Connor EF, Henderson JI, Friendrich SK, Grubbs RH, Bansleben DA (2000) Science 287:460
117. Saito J, Mitani M, Mohri J, Yoshida Y, Matsui S, Ishii S, Kojoh S, Kashiwa N, Fujita T (2001) Angew Chem Int Ed 40:2918
118. Saito J, Mitani M, Matsui S, Kashiwa N, Fujita T (2000) Macromol Rapid Commun 21:1333
119. Kashiwa N, Kojoh S, Kawahara N, Matsuo S, Kaneko H, Matsugi T (2003) Macromol Symp 201:319
120. Kawahara N, Saito J, Matsuo S, Kaneko H, Matsugi T, Kojoh S, Kashiwa N (2007) Polym Bull 59:177
121. Karavia V, Deimede V, Kallitsis JK (2004) J Macromol Sci A Pure and Appl Chem 41:115
122. Shiono T, Akino Y, Soga K (1994) Macromolecules 27:6229
123. Furuyama R, Mitani M, Mohri J, Mori R, Tanaka H, Fujita T (2005) Macromolecules 38:1546
124. Kojoh S, Matsugi T, Saito J, Mitani M, Fujita T, Kashiwa N (2001) Chem Lett 30:822
125. Saito J, Mitani M, Mohri J, Yoshida Y, Matsui S, Ishii S (2001) Angew Chem 113: 3002
126. Matsugi T, Matsui S, Kojoh S, Takagi Y, Inoue Y, Nakano T, Fujita T, Kashiwa N (2002) Macromolecules 35:4880
127. Mitani M, Mohri J, Yoshida Y, Saito J, Ishii S, Tsuru K, Matsui S, Furuyama R, Nakano T, Tanaka H, Kojoh S, Matsugi T, Kashiwa N, Fujita T (2002) J Am Chem Soc 124:3327
128. Weiser MS, Thomann Y, Heinz LC, Pasch H, Mülhaupt R (2006) Polymer 47:4505
129. Ono SS, Matsugi T, Matsuoka O, Kojoh S, Fujita T, Kashiwa N, Yamamoto S (2003) Chem Lett 32:1182
130. Ruokolainen J, Mezzenga R, Fedrickson GH, Kramer EJ, Hustad PD, Coates GW (2005) Macromolecules 38:851
131. Tian J, Hustad PD, Coates GW (2001) J Am Chem Soc 123:5134
132. Mori H, Yamahiro M, Tashino K, Ohnishi K, Nitta K, Terano M (1995) Macromol Rapid Comm 16:247
133. Mori H, Saito H, Terano M (1998) Macromol Chem Phys 199:55
134. Mori H, Yamahiro M, Prokhorov VV, Nitta K, Terano M (1999) Macromolecules 32:6008
135. Yamahiro M, Mori H, Nitta K, Terano M (1999) Polymer 40:5265
136. Li XF, Dai KD, Ye WP, Pan L, Li YS (2004) Organometallics 23:1223
137. Yoshida Y, Mohri J, Ishii S, Mitani M, Saito J, Matsui S, Makio H, Nakano T, Tanaka H, Onda M, Yamamoto Y, Mizuno A, Fujita T (2004) J Am Chem Soc 126:12023
138. Weiser MS, Mülhaupt R (2006) Macromol Rapid Commun 27:1009
139. Busico V, Cipullo R, Friederichs N, Ronca S, Togrou M (2002) Macromolecules 36:3806
140. Desurmont G, Tokimitsu T, Yasuda H (2000) Macromolecules 33:7679

141. Yasuda H, Furo M, Yamamoto H, Nakamura A, Miyake S, Kibino N (1992) Macromolecules 25:5115
142. Gromada J, Chenal T, Mortreux A, Leising F, Carpentier JF (2002) J Mol Cat A Chem 182/183:525
143. Desurmont G, Li Y, Yasuda H, Maruo T, Kanehisa N, Kai Y (2000) Organomet 19:1811
144. Killian CM, Tempel DJ, Johnson LK, Brookhart M (1996) J Am Chem Soc 118:11664
145. Doi Y, Ueki S (1982) Macromol Rapid Comm 3:225
146. Fukui Y, Murata M (2002) Appl Catal A General 237:1
147. Arriola DJ, Carnahan EM, Hustad PD, Kuhlman RL, Wenzel TT (2006) Science 312:714
148. Matsuo S, Matsugi T, Saito J, Kawahara N, Kaneko H, Kashiwa N (2006) Stud Surf Sci and Cat 161:1

Adv Polym Sci (2008) 217: 121–183
DOI 10.1007/12_2008_157

Published online: 2 July 2008

Synthesis and Properties of Cyclic Polymers

Kiyoshi Endo

Department of Applied Chemistry & Bioengineering, Graduate School of Engineering, Osaka City University, Sugimoto, Sumiyoshi-ku, Osaka 558-8585, Japan
endo@a-chem.eng.osaka-cu.ac.jp

Abstract Thousands of polymeric materials have been made into synthetic polymers, based on a linear structure, and used in commercial applications. The study of synthetic polymeric materials has focused on those derived from long chain linear molecules. Alternatively, cyclic polymers (also referred to as polymer rings or macrocycles) can be prepared, which not only can be branched or cross-linked, but can also form noncovalently linked structures based on their loop topology. Through a number of different approaches and advances in cyclization techniques, a wide range of novel cyclic polymers have been synthesized in good yields. This review will focus on a variety of synthetic methods and some properties of cyclic polymers using many polymerization mechanisms in various fields of polymer synthesis.

Keywords Cyclic polymer · Chain end functionalization · End-to-end cyclization reactions · Living polymerization · Ring-chain equilibrium

Abbreviations

β-BL	β-Butyrolactone
ε-CL	ε-Caprolactone
ATRP	Atom transfer radical polymerization
CDP	Cyclodepolymerization
CEVE	2-Chloroethyl vinyl ether
DMAc	Dimethyl acetoamide
DMF	Dimethylformamide
DMSO	Dimethyl sulfoxide
GPC	Gel permeation choromatography
HPLC	Liquid chromatography
M_w	Weight-average molecular weight
MALDI-TOF	Matrix-assisted laser adsorption/ionization time-of-flight
MWD	Molecular weight distribution
NIPAM	*N*-Isopropylacrylamide
NMP	Nitroxide mediated polymerization
P2VP	Poly(2-vinylpyridine)
PD	Degree of polymerization
PET	Polyethylene telephtalete
PMDETA	*N,N,N′,N″,N″*-Pentamethyldiethylenetriamine
RAFT	Reversible addition-fragmentation chain transfer
ROMP	Ring-opening metathesis polymerization
s-Bu	Secondary-butyl
SEC	Size exclusion choromatography
t-Bu	tert-Butyl
TEMPO	2,2,6,6-Tetramethyl-1-piperidinyl oxyl
THF	Tetrahydofuran
TMEDA	*N,N,N′,N*-Tetramethylethylenediamine
TTSBI	5,5′-Tetrahydroxy-3,3,3′,3′-tetramethyl spirobisindane

1 Introduction

Before the concept of macromolecules concept was established, most chemists believed that polymers consisted of aggregation of cyclic polymers [1]. After the concept was established, macromolecules became commonly represented as long, flexible random coiled chains, with the possibility of some cyclic structures, though in small amounts. Now various kinds of polymer structures have been architected by developing synthetic methods as depicted in Fig. 1. Many researchers have developed synthetic methods to control polymer architecture and tailor a material's properties for a specific application such as hyperbranched perfluorinated/hydrophilic block copolymers for antifouling coatings [2], branched and dendric structures for drug delivery [3, 4], globular polymers for the microelectronics industry [5], dendrimers for light harvesting applications [6, 7], and a variety of polymer architectures for catalysis [8, 9]. Alternatively, synthesis of cyclic polymers is also focused not only on novel synthetic methods but also on the characteristic properties that are different from linear polymers [10].

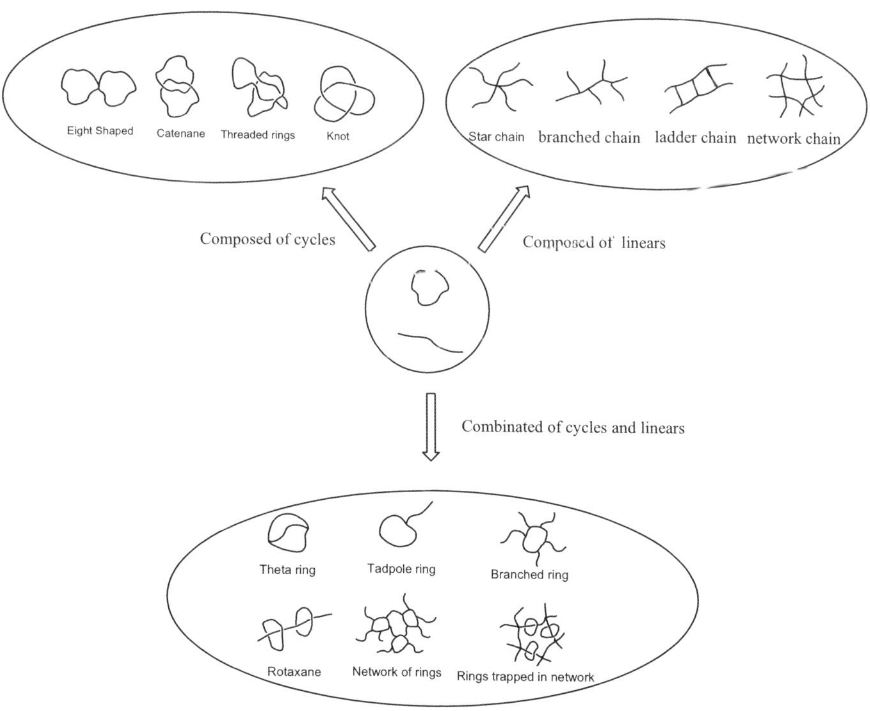

Fig. 1 Possible structures of linear chains and cycles

In recent years, cyclic polymers (also referred to as polymer rings or macrocycles) became easier to prepare. By a number of different approaches and advances in cyclization techniques, a wide range of novel cyclic polymers have been prepared in good yields [10]. In contrast to linear polymers, cyclic polymers are topologically distinct species, and all monomer units of cyclic polymers are chemically and physically equivalent. This equivalence is due to the fact that their properties are not affected by the nature of the end groups, since cyclic polymers have no chain ends. They include the radius of gyration, intrinsic viscosity, translational friction coefficient, critical solution temperature, refractive index, density, dipole moment, glass transition temperature, and surface property [11].

Another aspect of cyclic polymers is biological materials. The importance of large cyclic structures in biological chemistry and molecular biology is emphasized by a wide coverage of circular DNA, cyclic peptides and cyclic oligosaccharides and polysaccharides. Examples of their uses such as ring-opening polymerization reactions give commercially important materials. [12]

The ring topology is the potential to form unique polymer structures. Like linear polymers, cyclic polymers not only can be branched or crosslinked, but also can form non-covalently linked structures based on their loop topology. These are referred to as topological polymers, including rotaxane, catenane, threaded rings, and rings threaded by network chains. Recently, much attention has been paid to how their particular properties not only differ from linear polymers, but also how they differ from a component of an interlocked polymer system, such as polycatenanes and polyrotaxanes.

2 Synthetic Methods of Cyclic Polymers

2.1 Classification of Cyclization Process

Although some interesting synthetic processes of cyclic polymers have been addressed, the cyclization principle can be classified into two main methods as indicated in Fig. 2. One is the utilization of the ring–chain equilibrium that occurs in many polycondensation and ring opening polymerization. Another is the end-to-end cyclization method that can be used for synthesizing cyclic polymers from α,ω-difunctional linear precursors. The ring-closure reaction by the end-to-end cyclization is further divided into intermolecular reaction and intramolecular reaction, i.e., bimolecular process and unimolecular process, respectively.

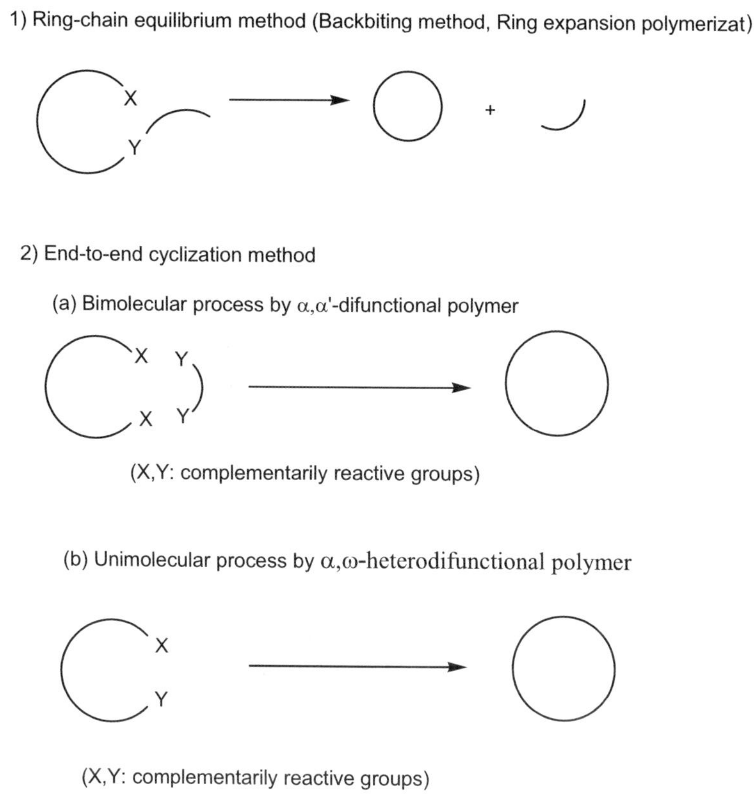

Fig. 2 Polymer cyclization processes

2.2 Ring–Chain Equilibrium Method

It is usually difficult to synthesize cyclic polymers by taking the random chain walk into consideration, but the formation of cyclic polymers is known to occur simultaneously with chain growth in many step-growth processes and even in addition polymerizations [13]. A theory that accounts for the competition between linear growth and cyclization was proposed by Jacobson and Stockmayer [14]. This theory has been shown to apply to ring-opening polymerization and step growth, which implies the presence of functional links along the polymer backbone. The ring content is enhanced upon dilution and the formation of small ring size is favored [14, 15]. The synthesis of such cyclic polymers has been reported since the 1950s in the preparation of polyesters [16, 17] and polyamides [18, 19]. Cyclic polymers that result from ring–chain equilibriums cover a broad range of molecular weights, because they mainly are formed by backbiting reactions. However, in the early days, experimental confirmation was much too difficult because the synthetic methods were not well developed. Today, the development of many analytical instruments allows separation of cyclic and linear constituents, and the ring–chain equilibrium has been utilized.

2.3
End-to-End Cyclization Method

End-to-end cyclization methods can be divided into bimolecular and unimolecular processes. One consists of bimolecular cyclization between a living α,ω-dicarbanionic polymer and a difunctional electrophile compound under extreme dilution conditions.

The controlled molecular weight can be synthesized by the end-to-end ring closure method. Reaction of a linear α,ω-difunctional chain results in a certain amount of cyclic polymers of the same degree of polymerization as that of the linear precursor. It follows that the lower the concentration of the cyclization medium was used, the higher the resulting cyclic polymer content was obtained. There is no theoretical limit as to the cyclic polymers that can be formed by this method.

A bimolecular process was reported by two independent groups, i.e., Höcker [19] and Rempp [20] in 1980. Macrocyclic polymers have been successfully prepared by the coupling reaction of a two-ended living polystyryl anion with a difunctional electrophile such as α, α'-dibromo-p-xylene under high dilution to yield cyclic and linear mixtures. The cyclic polymer was isolated by a fractional precipitation. This bimolecular end-to-end reaction process has been used for synthesizing cyclic polystyrene by many researchers [21–25].

Another strategy for cyclic polymers is an intramolecular (unimolecular) cyclization of linear α,ω-heterodifunctional polymers, which is proposed by Deffieux. In this case, the heterodifunctional polymers were used as precursors of the cyclic polymers, and α,ω-difunctional polymers should be designed before the cyclization [26–30]. The yields of the cyclic polymers were reported to be high (up to 90%).

3
Synthesis of Cyclic Polymers Using Addition Polymerization

3.1
Precursor Synthesis by Radical Polymerization

Recently, living radical polymerizations have been well developed, and various methods such as (1) iniferter mediated radical polymerization [31], (2) transition metal-mediated radical polymerization or atom transfer radical polymerization (ATRP) [32–34] (3) nitroxide-mediated free-radical polymerization (NMP) [35] and (4) reversible addition fragmentation transfer (RAFT) radical polymerization [36] have been used for preparing vinylic polymers with well-defined structures at both chain ends (Fig. 3). These linear polymers will be used for synthesizing cyclic polymers by the reaction of both chain ends, if both chain ends are designed for cyclization reaction.

Deffieux et al. have already prepared α-styrenyl-ω-acetal heterodifunctional vinylic polymers using living anionic polymerizations [26–30]. The heterotelechelic polystyrene chains containing α-hydroxy-ω-carboxy end groups using free radical polymerization were also prepared, and the intramolecular cyclization (unimolecular process) was examined [37]. The

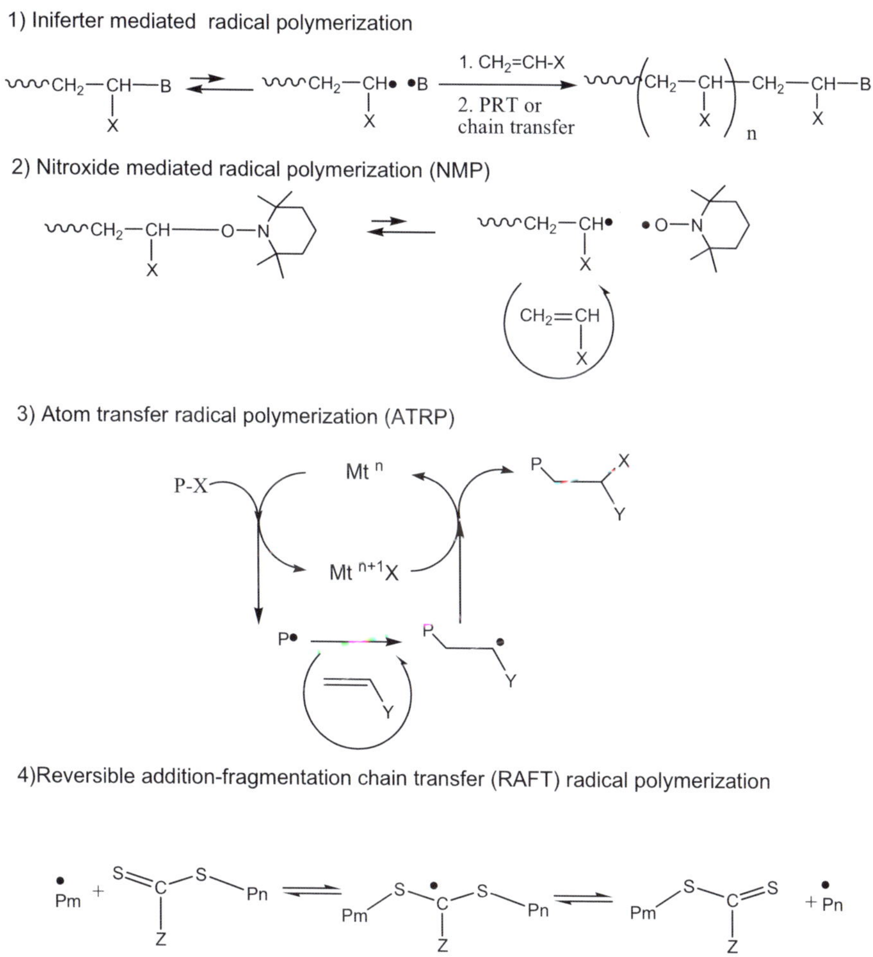

Fig. 3 Living radical polymerizations

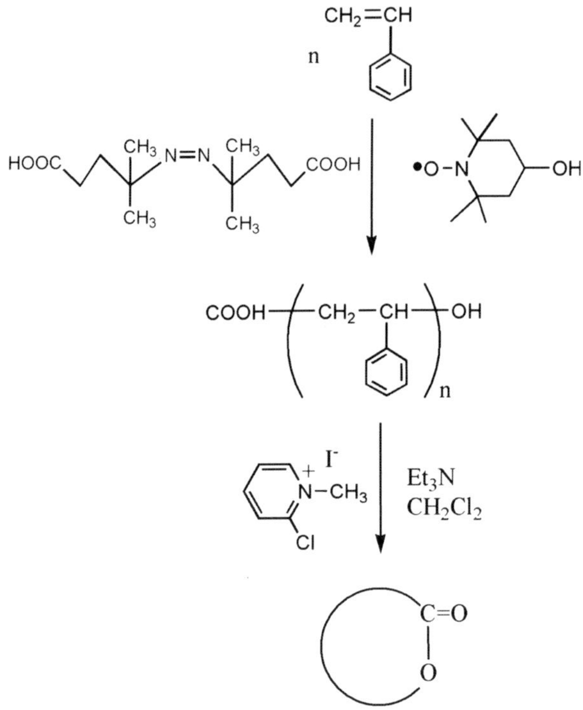

Fig. 4 Reaction scheme for synthesizing cyclic polystyrene using controlled radical polymerization

reaction process for preparing cyclic polymer is shown in Fig. 4. The controlled free radical polymerization of styrene was carried out using 4,4′-azobis(4-cyanovaleric acid) as the initiator and 4-hydroxy-TEMPO as the terminator, to give α,ω-heterodifunctional macromolecules ($M_n = 10^3 \sim 10^6$). The polymer was confirmed to be the cyclic structure by matrix-assisted laser desorption/ionization time-of-flight mass (MALDI-TOF mass) spectrometry and liquid chromatography (HPLC) at the exclusion–adsorption transition point analyses. The yield of macrocycles decreased due to additional styrene thermal initiation at 125 °C. The macrocyclic polystyrenes with controlled dimensions and narrow distribution have been obtained using "living" radical polymerization. The advantage of this method is that the linear precursor is prepared in one step via NMP with simple experimental conditions (purification, high-vacuum techniques) compared to ionic polymerization processes. The unimolecular cyclization was performed by esterification reaction to give cyclic polystyrene in a good yield.

Cyclic poly(methyl acrylate) with controlled ring size and its narrow distribution was synthesized by the ^{60}Co γ-ray-induced polymerization of methyl acrylate at – 30 °C in the presence of cyclic initiator without specific purification procedure as depicted in Fig. 5 [38]. The key point for synthe-

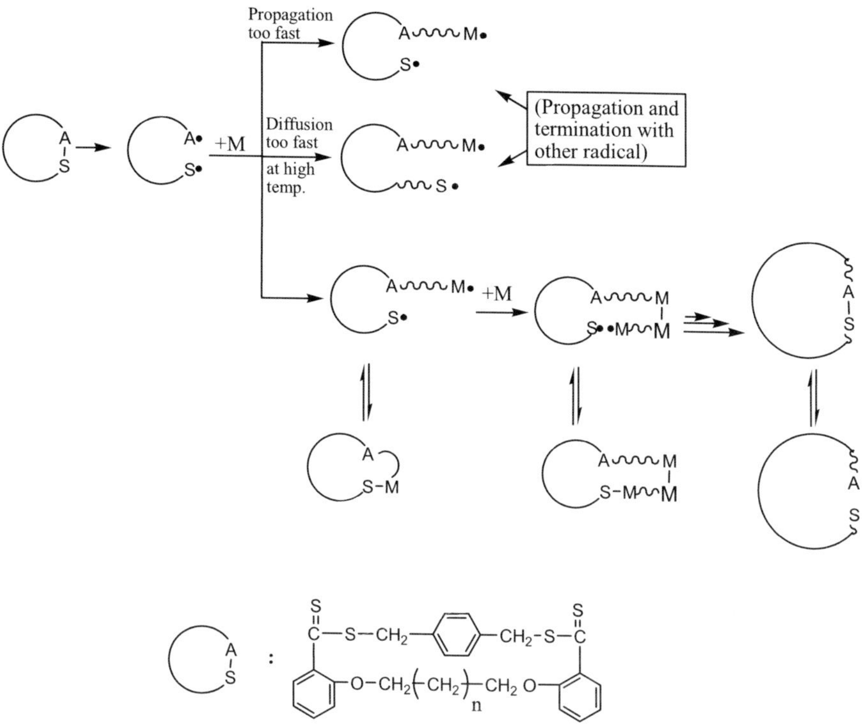

Fig. 5 Synthesis of cyclic polymer using polymerization of styrene with cyclic initiator

sizing the cyclic polymers is the reduced diffusion rate. The polymerization mechanism was proposed via repetitive monomer insertion into the cyclic initiator. The ring size was controlled by the ratio of monomer to initiator, and its distribution was narrow. With the same method and using cyclic poly(methyl acrylate) as macroinitiator instead of cyclic initiator, the amphiphilic cyclic block copolymer of poly(*N*-isopropylacrylamide-*block*-methyl acrylate-*block*-*N*-isopropyl acry-lamide) was also synthesized.

MALDI-TOF mass spectrometry analysis of poly(methyl acrylate) prepared by the free-radical polymerization of methyl acrylate (MA) in the presence of a cyclic dixanthate under γ-ray irradiation revealed that there are at least three distributions, i.e., molecular mass for $[1\text{-}(MA)_n\text{-}H]^+$ of cyclic polymers, $[1\text{-}(MA)_n\text{-}THF\text{-}H]^+$, and $[1\text{-}(MA)_n\text{-}(THF)_2\text{-}H]^+$ of linear polymers were observed. The relative content of the cyclic polymers markedly increases at a lower temperature, which may be related to the reduced diffusion rate and the suppressed chain-transfer reaction at the low reaction temperature [39].

"Click" chemistry has been used extensively since its introduction in organic chemistry, due to the high efficiency and technical simplicity of the reaction [40]. The most popular "click" reaction has been the copper-catalyzed dipolar cycloaddition of a terminal alkyne and an azide to form

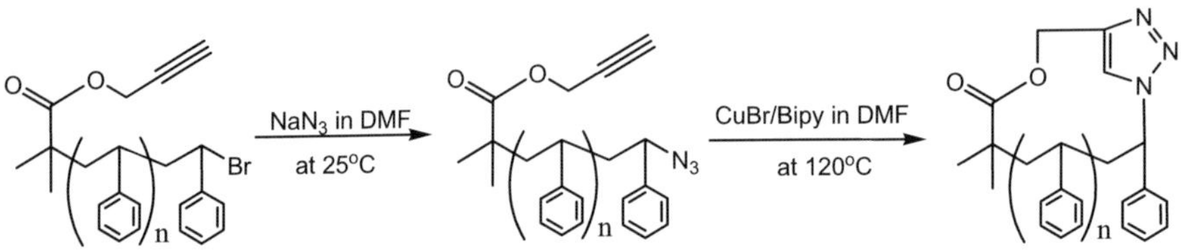

Fig. 6 The terminal azidation and "Click" cyclization of polystyrene prepared via ATRP

a 1,4-disubstituted 1,2,3-triazole [41]. Recently, Boyd et al. reported a new route for synthesizing well-defined macrocyclic polymers using "click" cyclization [42]. The linear polystyrene precursors were prepared using the standard ATRP techniques, i.e., propargyl 2-bromoisobutyrate was used as initiator with Cu(I)Br and *N,N,N′,N″,N″*-pentamethyldiethylenetriamine (PMDETA) as catalyst. The reaction was carried out in bulk to consistently yield polymers with a polydispersity of less than 1.2. The bromine-terminated product was purified by extraction from water into CH_2Cl_2 and precipitation into methanol. Azidation of the end group was carried out in DMF with sodium azide (Fig. 6). The high efficiency of the "click" reaction enables the cyclization of styrenic polymers prepared by ATRP polymerization. Both the end-group modification and the cyclization of the linear polystyrene appear to be nearly quantitative.

Poly(*N*-isopropylacrylamide) (PNIPAM) is considered by many chemists as the prototype of such "smart" polymers. The conformational change of PNIPAM in water has motivated many fundamental investigations aimed at understanding the physics of the coil-to-globule transition. For this purpose, the cyclic PNIPAMs was synthesized by ring closure of α,ω-heterodifunctional telechelic PNIPAM precursors produced by the RAFT polymerization process, following the strategy reported recently for the synthesis of cyclic polystyrenes based on the copper(I)-catalyzed 1,3-dipolar azide to alkyne cycloaddition or "click chemistry" [43]. To obtain telechelic PNIPAM suitable for "click" ligation, an azide and a propargyl group were introduced on the α- and ω-ends of each PNIPAM chain, which was accomplished in three steps as shown in Fig. 7. For cyclic/linear pairs of identical molecular weight, the following results were obtained: (1) the endotherm is broader for solutions of *cyclic*-PNIPAM; (2) the maximum of the endotherm (TM) of *cyclic*-PNIPAM solutions is shifted to higher temperature by $\sim 5\,^{\circ}C$; (3) the enthalpy of the transition per NIPAM unit (ΔH) is smaller for *cyclic*-PNIPAM, and the value decreases with decreasing ring size, whereas in the case of *linear*-PNIPAMs, ΔH is independent of chain length; (4) the transmittance of *cyclic*-PNIPAM solutions decreases gradually upon heating, whereas solutions of *linear*-PNIPAM undergo a sharp decrease in transmittance. Recently, similar thermoresponsive water-soluble *cyclic*-PNIPAM was found via intramolecular "click" cyclization of α-alkyne-ω-azido heterodifunctional PNIPAM linear precursor (*linear*-PNIPAM-N3), which was prepared by the

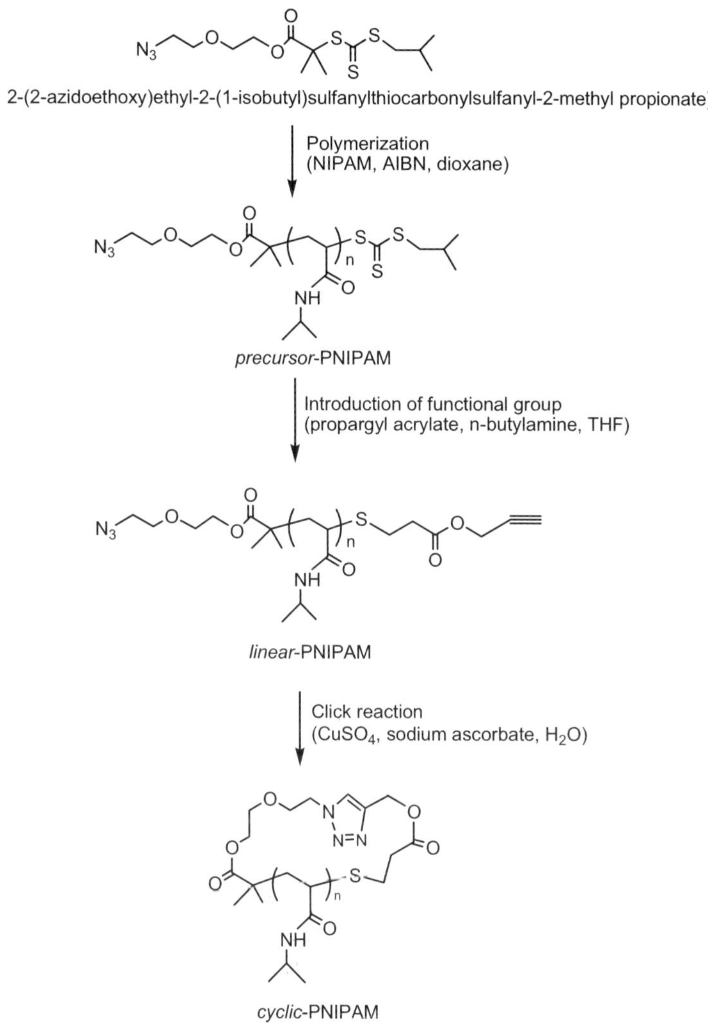

Fig. 7 Synthesis of cyclic PNIPAM using Click reaction

ATRP of *N*-isopropylacrylamide (NIPAM) and subsequent nucleophilic substitution reaction with NaN_3 [44].

Synthesis of monocyclic polystyrene using the reversible coupling/cleavage of thiol/disulfide groups was reported by Whittaker et al. [45]. The α,ω-dithiol polystyrene was synthesized by polymerizing styrene in the presence of a difunctional RAFT agent and subsequent conversion of the dithioester end groups to thiols via the addition of hexylamine. Oxidation gave monocyclic polymer chains, but linear multiblock polymers were also formed at high polymer concentrations as depicted in Fig. 8. To control the molecular

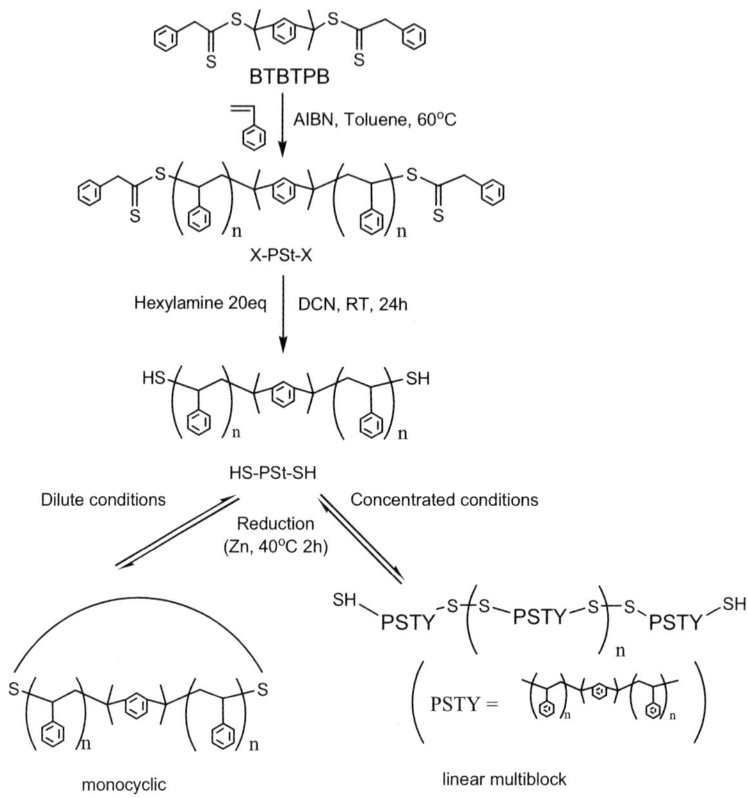

Fig. 8 Synthesis of monocyclic (one disulfide linkage) and linear multiblock polymers through thiol–thiol coupling reactions

weight distribution (MWD) of these linear multiblocks, monofunctional prepolymer was added. Disulfide linkages can be readily reduced back to the starting polystyrene with the thiol end groups, which have potential use as a recyclable polymer material.

3.2 Precursor Synthesis by Cationic Polymerization

Diffiuex investigated a synthesis of cyclic poly(vinyl ether) using cationic polymerization [26,28]. The reaction process is depicted in Fig. 9. They studied on the living cationic polymerization of 2-chloroethyl vinyl ether (CEVE) initiated with the HI adduct of 4-(vinylbenzyloxy)butyl vinyl ether prepared by reacting chloromethyl styrene with sodium salt of 4-hydroxybutyl vinyl ether in THF at 80 °C. By the cationic polymerization of CEVE, α,ω-heterofunctional linear polymer precursor of cyclic poly(CEVE) was produced. The MWDs of the polymers were unimodal and very narrow (< 1.2),

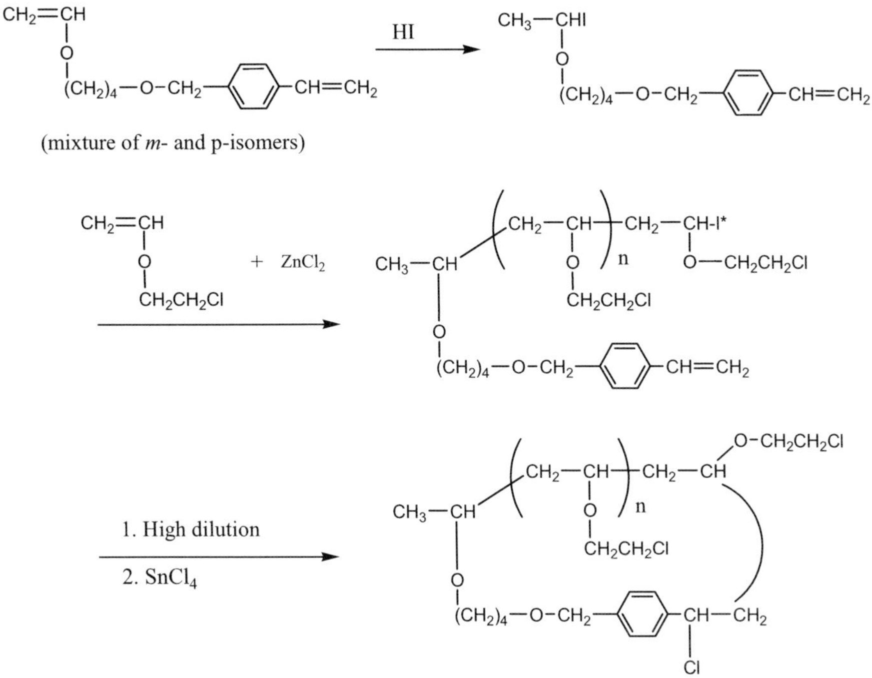

Fig. 9 Synthesis of cyclic poly(CEVE) in the presence of $SnCl_4$

indicating that the polymerization proceeds by a living cationic mechanism. The cyclization of the linear polymer was carried out in the presence of $SnCl_4$ under high dilution conditions. The cyclic polymer was then obtained in a high yield. They also synthesized the cyclic block copolymer consisting of styrene and CEVE units through a similar process [28].

3.3 Precursor Synthesis by Anionic Polymerization

3.3.1 Bimolecular Cyclization of α, α'-Difunctional Precursor

Szwarc, Levy, and Milkovich first demonstrated the "living" nature of the polymerization of styrene initiated with sodium naphthalene to give a bifunctional living chain end [46]. It will be possible to synthesize cyclic polymers using a bifunctional living chain end. This basic strategy was suggested some time ago with the discovery of living polymer of styrene. It is used to form a narrow molecular weight distribution polymer with two carbanion ends and react with a difunctional electrophile under an extreme dilution condition [47].

Höcker reported a pioneering work on the synthesis of cyclic polymer using living anionic polystyrene. Macrocyclic polystyrene samples ($30 < DP < 250$) with narrow molecular weight distribution was synthesized by living bifunctional anionic polymerization initiated with sodium naphthalene and by coupling with *α*,*α*′-dichloro-*p*-xylene as a bifunctional terminating agent [48]. Linear and cyclic molecules were separated by fractionation after the reaction of the linear molecules having chlorine end groups with high molecular weight living polystyrene. The ratio of the intrinsic viscosity of ring and linear polymer chains was close to 0.65 in cyclohexane at 34.5 °C. On the other hand, Rempp et al. used potassium naphthenide in a mixture of benzene and THF to polymerize styrene and *α*,*α*′-dibromo-*p*-xylene as the cyclization agent [49, 50]. Vollmert et al. used sodium naphthenide in THF to polymerize styrene and *α*,*α*′-dibromo-*p*-xylene to perform ring closure [51].

Hogen-Esch reported the glass transition temperatures of linear and macrocyclic poly(2-vinylpyridine) (P2VP) and polystyrene [52]. However, the P2VP macrocycles were synthesized by reaction of the lithium dianion "living" precursors with 1,4-bis(bromomethyl)benzene (EBr_2) as shown in Fig. 10. Macrocyclic polystyrene was synthesized by initiation of styrene in THF at about – 70 °C using lithium naphthalene as initiator. For both the P2VP and polystyrene cases the difference in glass transition temperature, ΔT_g between linear and macrocyclic polymers increases with decreasing DP. It is unclear to what degree this increase in ΔT_g is due to the absence of end groups in the macrocycles and to the increasing stiffness of the macrocycles as the DP decreases.

Alternatively, P2VP with cyclic structure was synthesized by the reaction of linear P2VP, having two living ends followed by coupling with a bifunctional coupling agent [53]. The crude product was fractionated by preparative size exclusion chromatography to enrich the cyclic polymer portion. The existence of the ring structure in the fractionated P2VP sample was directly

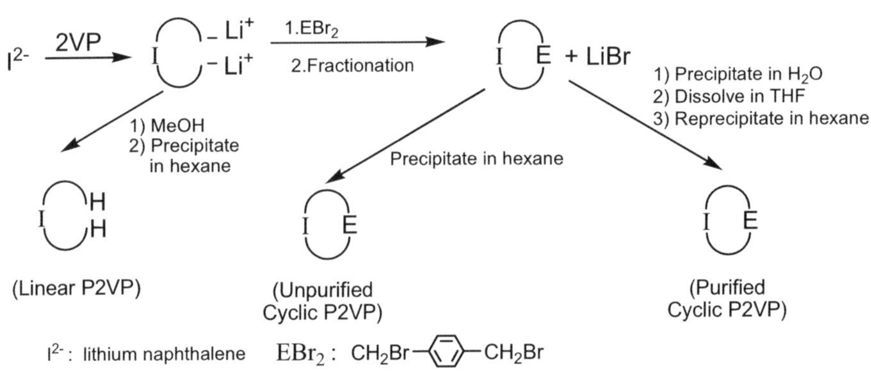

Fig. 10 Synthesis of Linear and Macrocyclic P2VP's and Purification of Cyclic P2VP

confirmed by pyrolysis-gas chromatography/mass spectrometry, and characteristic fragments were observed.

Transmission electron microscopy and small-angle X-ray scattering were used to study the effect of looped vs. bridged chain conformations on the morphological characteristics of microphase-separated cyclic block copolymers prepared by end-coupling the corresponding linear triblock precursors with 1,4-bis(bromomethyl) benzene and Me_2SiCl_2, respectively. For a given cyclic diblock/linear triblock pair, the cyclic polymer always had a smaller domain spacing [54, 55]. Macrocyclic polystyrenes were also prepared by coupling a two-ended living precursor dianions with 1,3-bis(1-phenylvinyl)benzene. Macrocycles were obtained with a yield ranged between 40 and 55% [56]. Simultaneously, besides the expected cyclic polymer, polycondensates (linear and cyclic) were formed. Macrocyclic products were separated from the residual linear precursor. The linear polycondensates byproducts were separated by preparative HPLC at the exclusion–adsorption transition point. Isolation of highly pure cyclic polystyrenes with different sizes was performed.

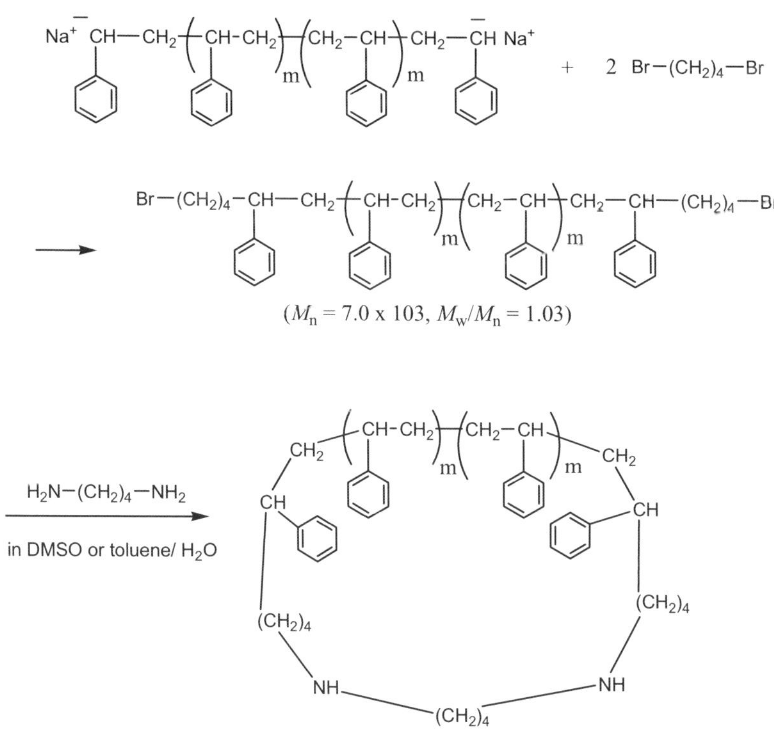

Fig. 11 Synthesis of cyclic polystyrene by interfacial condensation

The effect of solvent on the synthesis of cyclic polyisoprene was investigated [57]. Cyclic polyisoprene has been synthesized by the reaction of α,ω-dilithiopolyisoprenes with dichlorodimethylsilane or 1,2-bis(isopropenyl-4-phenyl)ethane. Using the dihalide compound, the effect of the solvent polarity on the coupling reaction is more pronounced in the case of α,ω-dilithiopolyisoprene than with monofunctional polyisoprenyllithium. The yield in cyclic compound falls from 88% in pure hexane down to 53% in the presence of 15 vol % of THF. Using the 1,2-bis(isopropenyl-4-phenyl)ethane as linking agent the addition of THF is required, but the formed cycle retains its living character and allows the synthesis of cycles having two arms (after addition of isoprene) and of a bicyclic structure after a second cyclization reaction.

On the other hand, Ishizu et al. [58] reported the synthesis of cyclic polystyrene using interfacial condensation reaction of α,ω-dibromopolystyrene prepared from living polystyrene initiated with sodium naphthalene and terminated with 1,4-dibromobutane and then tetramethylenediamine as depicted in Fig. 11. The reaction was carried out in organic solvent/water to yield in more than 90%. The effect of solvent on the yield of cyclic polymer was observed, and the yield of cyclic product obtained in DMSO was higher than that in toluene. Since DMSO dissolves in both water and toluene, the reaction proceeded faster than that in toluene.

Recently, cyclic diblock copolymers of styrene and butadiene were synthesized (Fig. 12). The synthetic approach of the cycles involved the reaction of (1,3-phenylene)bis(3-methyl-1-phenyl-pentylidene)dilithium initiator

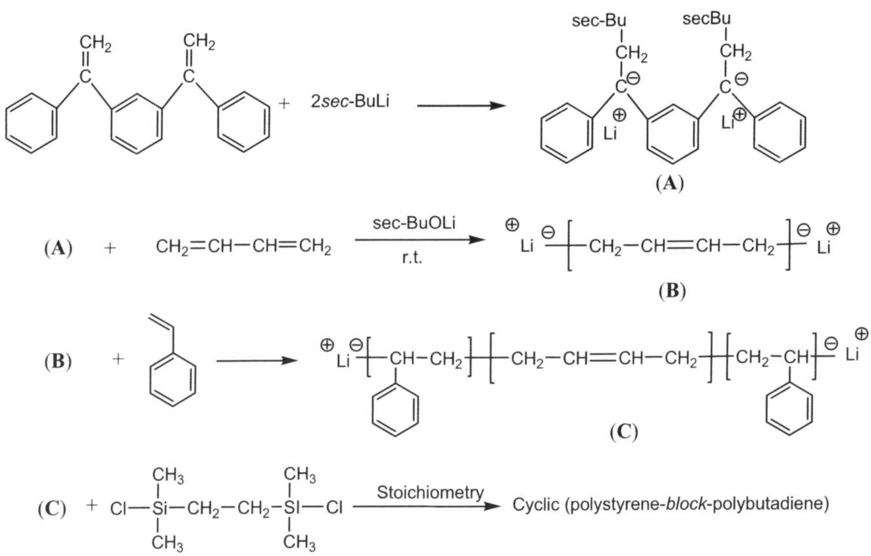

Fig. 12 Synthesis cyclic block copolymer of styrene and butadiene using bimolecular cyclization

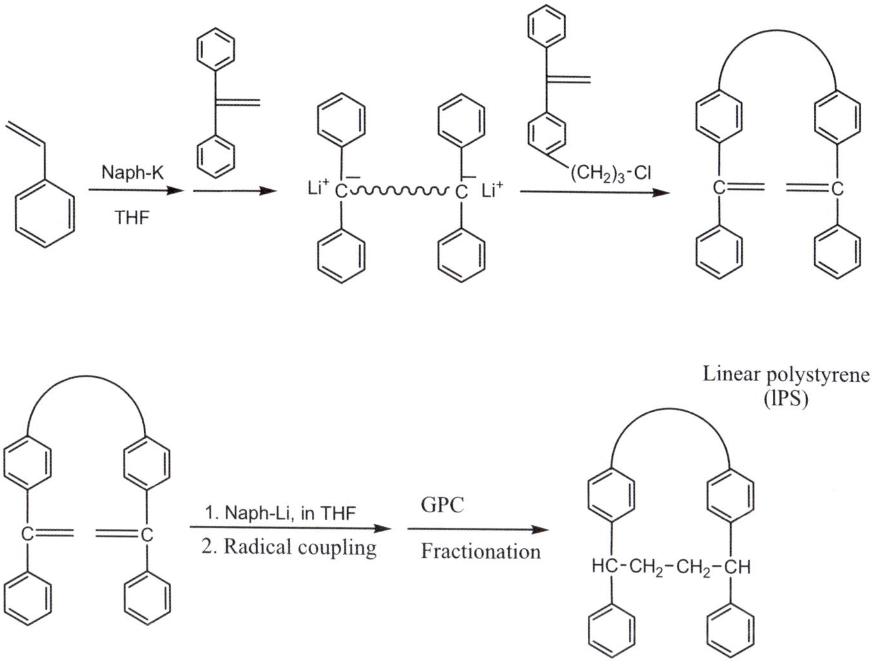

Fig. 13 Synthesis of cyclic polystyrene by radical coupling reaction

with butadiene in the presence of *sec*-BuOLi, followed by polymerization of styrene. The cyclization of the resulting α,ω-difunctional triblock copolymer was performed by using bis(dimethylchlorosilyl)-ethane, under high dilution conditions [59].

A synthetic route of cyclic polystyrene, as depicted in Fig. 13, was reported [60]. A polystyrene precursor with two end vinyl groups and potassium naphthalenide as a coupling agent was synthesized. Cyclic polystyrenes were then prepared by the novel radical coupling agent with molar masses up to 600 kg/mol. From the SEC-fractionation, the higher molecular weight polycondensates were removed completely and obtained an amount of highly pure single chain cyclic polystyrenes. With the HPLC techniques, not only the linear polycondensation molecules but also the cyclic polycondensates were successfully separated. From the careful analysis of the chromatograms obtained, it has been clarified that the product obtained from the end-to-end ring closure reaction of a telechelic polystyrene under an extremely diluted condition includes relatively high abundance of dimeric and trimeric molecules, as well as a quite high yield of the of the monomeric one [61].

Polystyrene-*block*-polyisoprene-*block*-polystyrene (SIS) with different compositions were synthesized and cyclized by the coupling reaction between two chain end groups on the same molecules as depicted in Fig. 14 [62]. From

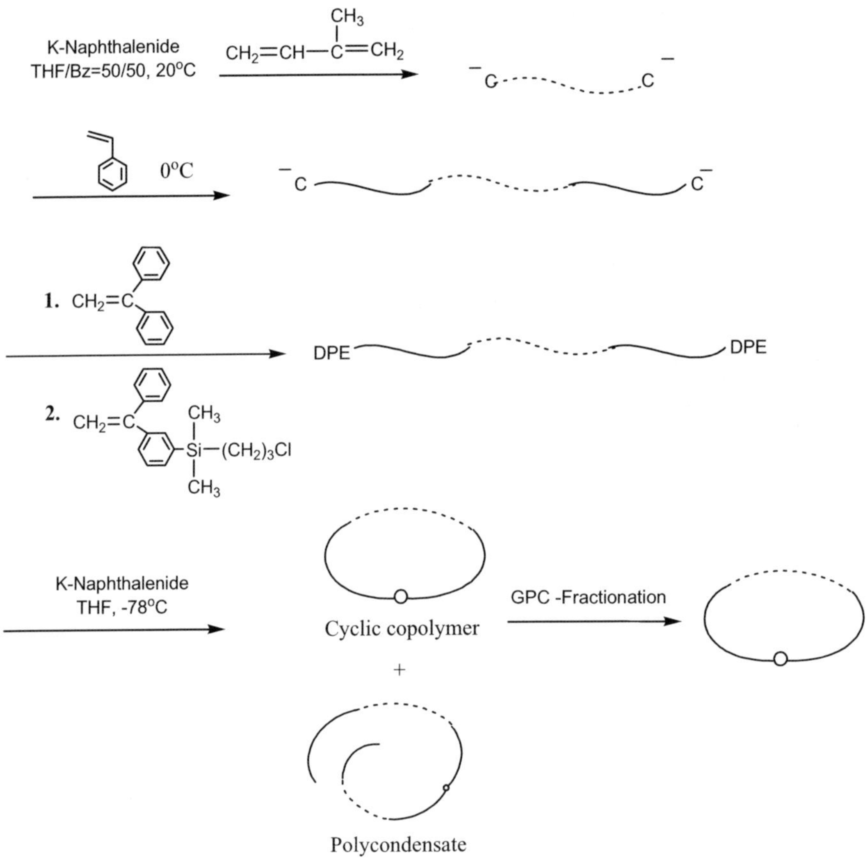

Fig. 14 Synthesis of cyclic poly(styrene-*block*-isoprene-*block*-styrene)

GPC fractionation and ozonisis of the products, the yield of cyclic polymer was estimated to be ca. 90%. The morphological transition of the cyclic block polymer depends on composition in essentially the same manner as that of the linear triblock copolymers, whereas the domain spacing of polystyrene-*block*-polyisoprene cyclic block copolymers were all smaller than those of the corresponding SIS linear triblock copolymers, which is attributed to looped chain conformation.

3.3.2
Unimolecular Cyclization of α,ω-Heterodifunctional Polymer

Deffieux et al. proposed a new approach to the synthesis of macrocyclic polymers with unimolecular end-to-end coupling reaction using α,ω-heterodifunctional linear polymer precursor [26]. In the method, which involves the bimolecular end-to-end ring closure of an anionically prepared α,α'-difunctional linear polymer using a coupling agent, low cyclization yields were sometimes observed generally due to polycondensation. In contrast to the bimolecular process, the unimolecular end-to-end ring closure of an anionically prepared α,ω-difunctional linear polymer has advantages, such as a single reaction process much less sensitive to the extremely low concentrations required for cyclizations. This new technique has been applied to the synthesis of cyclic polystyrenes in high yields (85–90%) [30].

The cyclic polystyrene was also prepared by a single-step, and pseudo-unimolecular cyclization of a linear α-(diethoxyethyl)-ω-styrenylpolystyrene. This cyclic precursor was synthesized by living anionic polymerization according to the reaction pathway given in Fig. 15. The cyclic polymer was analyzed by liquid chromatography at the critical point of adsorption [63]. Using silica gel as the stationary phase and THF-hexane as the eluent, the macrocyclic oligomers are separated from their linear precursors. The cyclization yield is 90–95%. Such cyclic polystyrene was analyzed by MALDI mass spectrometry, which was a new and powerful technique for the analysis of macrocyclic polymers [64].

Roovers also synthesized high molecular weight cyclic polystyrene (5×10^3 to 4.5×10^5) with a similar process [65]. The α,ω-heterodifunctional product prepared from the reaction of two-ended living polystyryl-sodium with

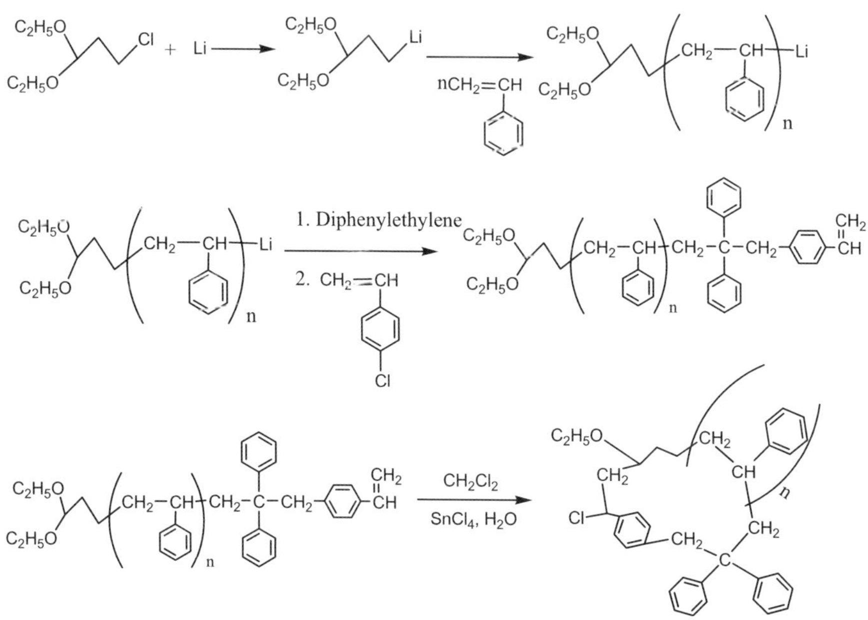

Fig. 15 Reaction pathway for the synthesis of cyclic polystyrene by unimolecular process

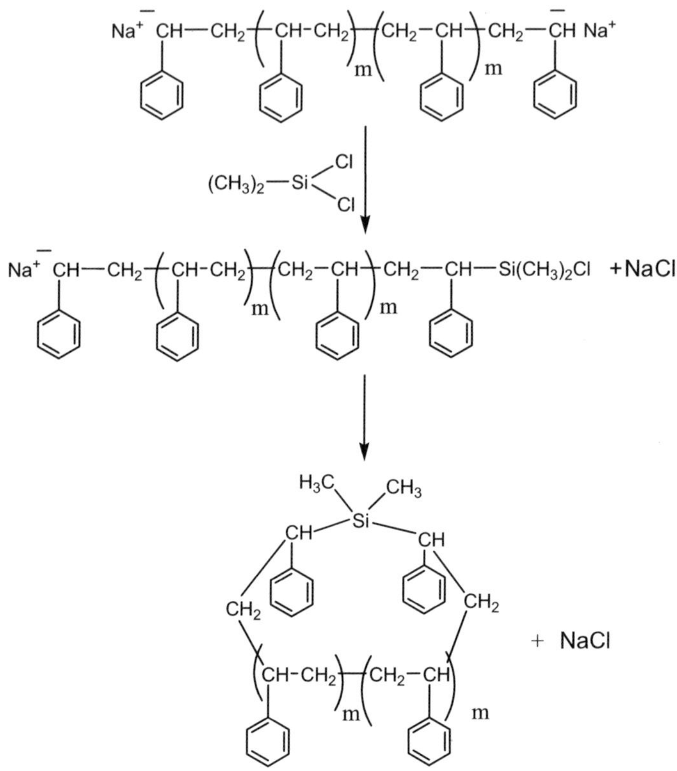

Fig. 16 Synthesis of cyclic polystyrene using dimethyldichlorosilane as the cyclization agent

dimethyldichlorosilane as the cyclization agent was used to prepare narrow molecular weight distribution cyclic polymers as shown in Fig. 16. The cyclic and linear polymers of the same molecular weight can be fractionally precipitated. Ultracentrifugation sedimentation and GPC were used to monitor the purity of the ring polymers. The purity of the very dilute anionic polymerization systems at the point of the cyclization reaction seems to be the limiting factor for the preparation of high molecular weight rings. Intermolecular polycondensation took place as a side reaction, leading to the chain extension of the polymer. Cyclic polybutadiene was also synthesized by a similar procedure [66].

A novel synthesis of α-carboxyl, ω-amino heterodifunctional polystyrene and its intramolecular cyclization was reported. Cyclic polystyrene with controlled size was synthesized by the intramolecular cyclization of linear α-carboxyl, ω-amino heterodifunctional polystyrene under high dilution in the presence of 2-chloro-1-methylpyridinium iodide [67]. The linear precursor polymer was derived from α-diethyl acetal, ω-amino heterodifunc-

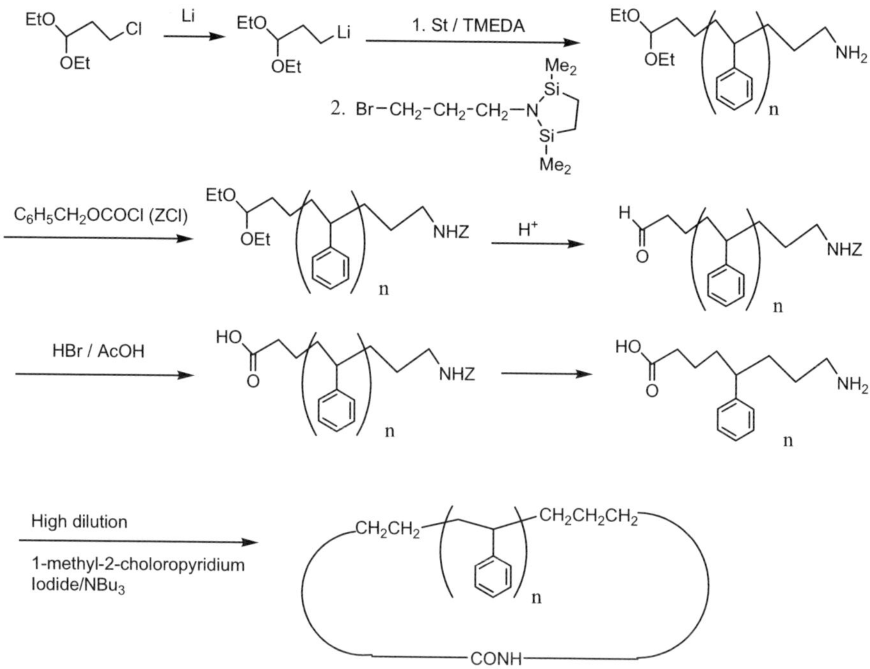

Fig. 17 Synthesis of cyclic polystyrene by amidation

tional polystyrene, which was initiated with 3-lithiopropionaldehyde diethyl acetal in benzene in the presence of *N*,*N*,*N′*,*N*-tetramethylethylenediamine (TMEDA) at room temperature and terminated with 2,2,5,5-tetramethyl-1-(3-bromopropyl)-1-aza-2,5-disilacyclo-pentane as an aminating agent, as shown in Fig. 17. The cyclic structure of the polymer was characterized by ^{1}H NMR, ^{13}C NMR, and IR spectra of the products.

The formed cyclic polystyrene was used to synthesize a well-defined cyclic macromonomer as shown in Fig. 18, and the radical copolymerizations with vinyl monomers were performed. The copolymerization with *t*-butyl acrylate in benzene at 70 °C gave an insoluble cross-linked polymer with a good swelling property [68]. The insoluble material was converted to soluble polymers by the cleavage of cyclic polystyrene from the backbone chain, suggesting a mechanically cross-linking process, due to the threading of the cyclic polystyrene by a segment of another polymer chain during the copolymerization [69].

A cyclic polymer of methyl methacrylate was also synthesized using a heterotelechelic polymer precursor, although the formation of living polymer of methyl methacrylate is more difficult than that of styrene. Mizawa et al. described the synthesis of heterotelechelic poly(methyl methacrylate) containing α-maleimide-ω-dienyl end groups and its intramolecular cyclization

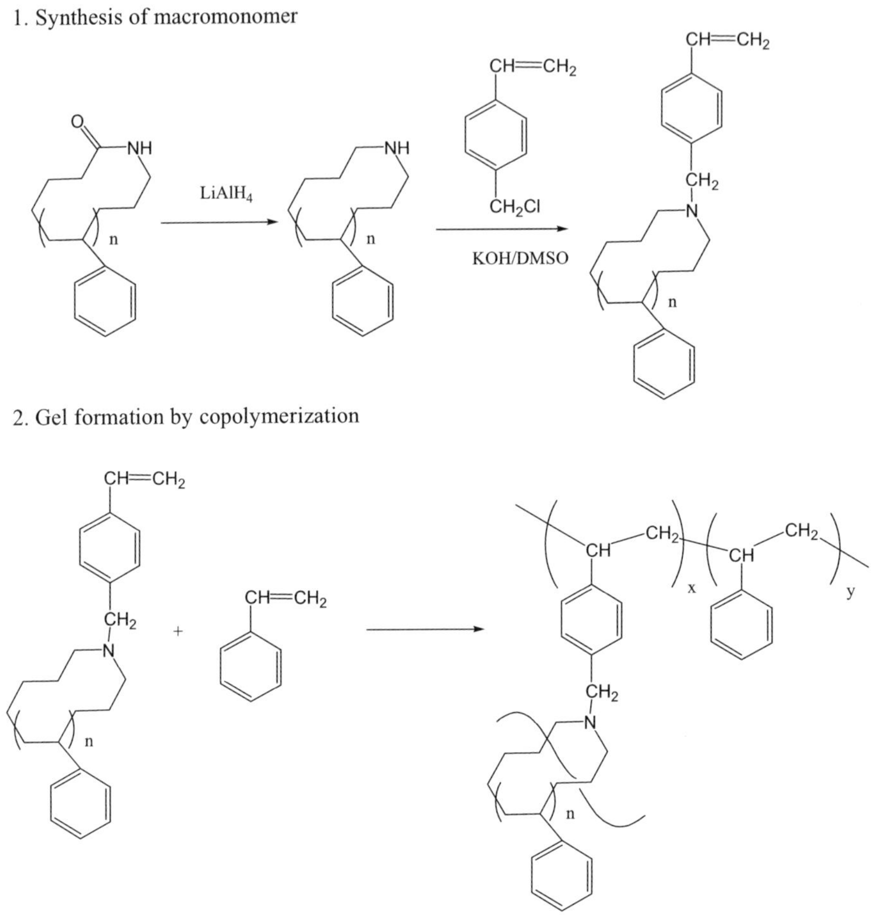

Fig. 18 Copolymerization of the cyclic polystyrene macromonomer and a vinyl monomer

via a Diels–Alder reaction [70]. Moreover, α-amino, ω-carboxyl heterodifunctional poly(methyl methacrylate) was prepared by a living anionic polymerization of methyl methacrylate using *N*,*N*′-diphenylethylenediamine monolithium amide and succinic anhydride as an initiator and terminator, respectively. A well-defined cyclic poly(methyl methacrylate) was synthesized by the intramolecular cyclization of the precursor [71].

The viscosity of a linear polyelectrolyte solution depends on the conformation of the molecules, which in turn is affected by intramolecular electrostatic interactions between charged segments located along the polymer backbone, but the interactions in systems of charged polyelectrolytes are still far from being understood. The study on the solution property of cyclic polyelectrolyte is of interest, since the chain expansion of a cyclic

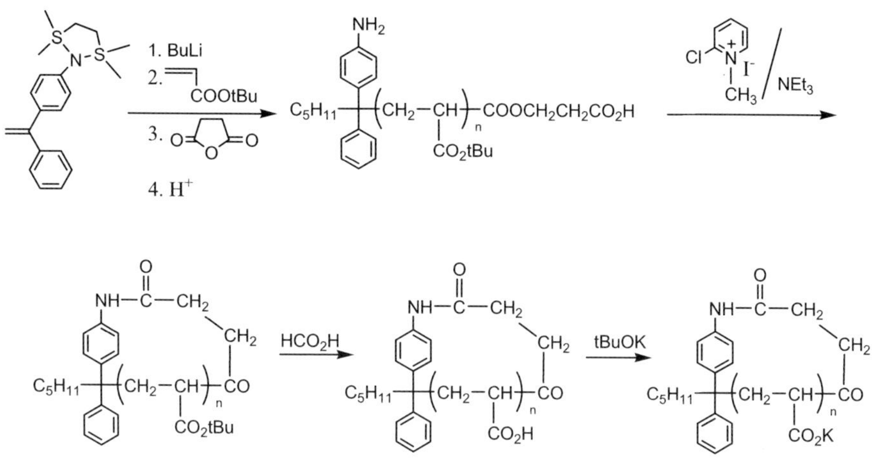

Fig. 19 Synthesis of cyclic potassium polyacrylate by amidation

polyelectrolyte with relatively low molecular weight is restricted due to the presence of topological restraint. For this purpose, a novel preparation of well-defined cyclic poly(acrylic acid) and cyclic potassium polyacrylate with relatively short chain length was synthesized, according to the process indicated in Fig. 19. The cyclic structure was confirmed by MALDI-TOF mass spectrometry. A well-defined cyclic poly(acrylic acid) and cyclic potassium polyacrylate, which represent interesting model polyelectrolytes, were synthesized [72].

3.4 Synthesis of Cyclic Polymers Using Acetylene Polymerization

Synthesis of linear and cyclic alkanes (C_nH_{2n+2}, C_nH_{2n}) with $24 \leq n \leq 288$ was achieved by the polymerization of α,ω-diacetylenic monomers with Cu catalysts followed by hydrogenation for evidence of chain folding, chain packing, and conformations [73, 74]. The reactions for preparation of cyclic polyethylene are shown in Fig. 20. Evidence of chain folding in long chain linear alkanes (C_nH_{2n+2} with $n \geq 168$) was obtained by Raman and Fourier-transform IR spectroscopy. Investigation of cyclic alkanes (C_nH_{2n} with $24 \leq n \leq 288$) using vibration spectroscopy allowed identification of specific bands due to folding as well as determination of conformational sequences in the fold region. The cyclic alkanes (C_nH_{2n} with $n = 48$, 72, and 96) were crystallized from solution with monoclinic structures, whereas the linear alkanes (C_nH_{2n+2} with $n = 144$ and 288) were obtained as orthorhombic crystals. The cyclic alkanes (C_nH_{2n} with $n = 72$ and 96) crystallized from the melt as crystallites with a high degree of orthorhombic character.

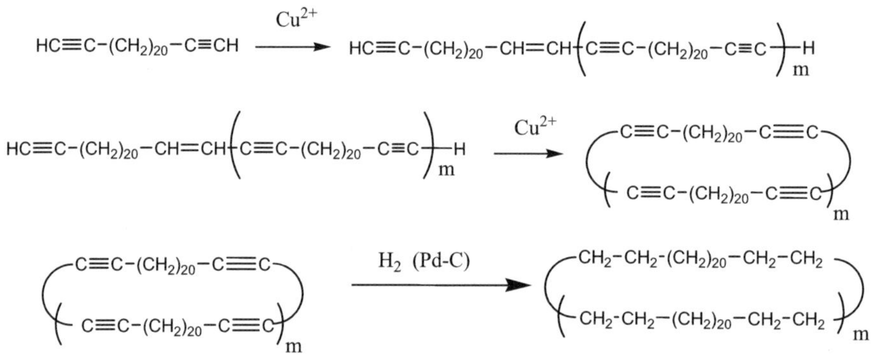

Fig. 20 Synthesis of cyclic polyethylene from acetylenic monomer

4
Synthesis of Cyclic Polymer by Polycondensation

In polycondensations, cyclization reaction takes place as a side reaction during the polymerization, so that low molecular weight cyclic oligomers can be formed under a dilute reaction condition in the polycondensation. Thus, many polycondensates exhibit ring–chain equilibriums, as is the case in polyester, polyamide, polysiloxanes, and polyacetals. This phenomenon, called backbiting, ring expansion, or reshuffling, is basically ring–chain equilibrium.

4.1
Cyclic Polyesters

Cyclic oligoesters from trimers to heptamers were synthesized by the reaction of oligo-dicarboxylic acids and oligo-diols in benzene under high dilute conditions as shown in Fig. 21 [75]. The cyclic trimers can be separated by extraction from high molecular weight PET (m.p. 325–327 °C) [16], and cyclic oligomer with higher molecular weight was also separated from PET. The amount of cyclic oligomers including ethylene glycol was ca. 1.3–1.7% [17].

Mass spectral data have been directly used to determine the distributions of cyclic oligomers formed during degradation reactions. Thermal degradation reactions occurred in polyesters to give cyclic oligomers. The distributions of cyclic oligomers allow discrimination between thermodynamically and kinetically controlled cyclization processes [76]. Figure 22 shows the different polycondensation processes to yield cyclic polyesters in preparation of macrocyclic poly(1,4-butane diol-ester)s by polycondensation of 2-stanna-1,3-dioxepan with dicarboxylic acid dichlorides [77] and by polycondensations of aliphatic dicarboxylic acid dichloride with catechol or bistrimethylsilyl catechol in bulk or concentrated solution [78]. Various diols

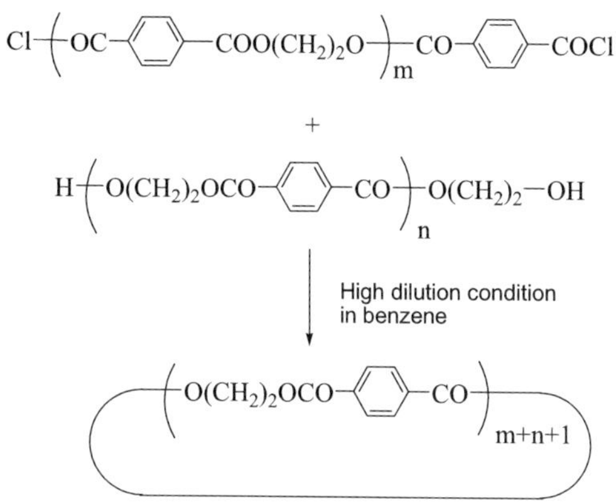

Fig. 21 Synthesis of cyclic oligoester by polycondensation

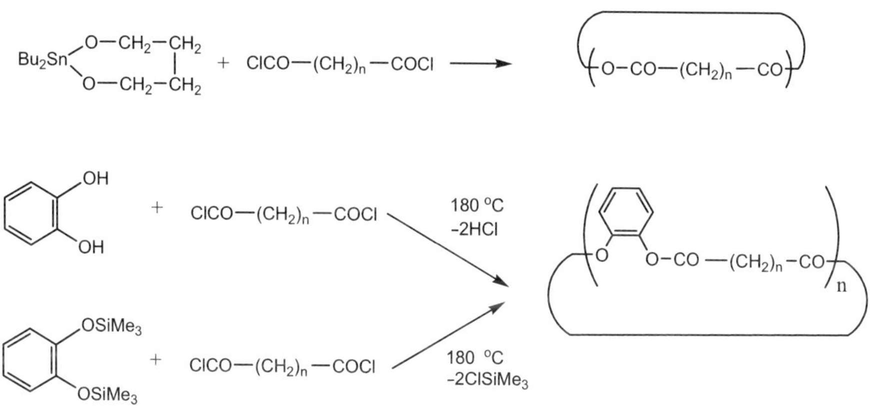

Fig. 22 Synthesis of cyclic polymers by polycondensation

were polycondensed with dicarboxylic dichlorides in the presence of pyridine or γ-picoline. Aliphatic dicarboxylic acid dichlorides were reacted with 1,4-butanediol, catechol resorcinol, methylhydroquinone or bisphenol-A [77–79]. The resulting polyesters were characterized by ^{1}H NMR spectroscopy and SEC measurements. From analysis of MALDI-TOF mass spectrometry, cyclic oligomers or polyesters were confirmed to be the main products in all cases. An excess of pyridine, long reaction times and a slight excess of the acid chloride play the key role for the success of the optimization. Cyclic oligomers and polyesters are the main reaction products regardless of the monomer structures. The cyclizations are an inherent and necessary component in step-growth polymerizations conducted under ideal conditions (no side reactions and equilibrations, perfect stoichiometry, and homogeneous phase), and the cyclizations are a decisive limitation of the chain growth even under ideal conditions. The cyclic products were obtained by increasing conversion, and almost cyclic polymers were obtained at a high polymer yield.

Tin-containing polylactones obtained from ring-expansion polymerizations of β-butyrolactone (β-BL) or ε-caprolactone (ε-CL) initiated with 2,2-dibutyl-2-stanna-1,3-dioxepane were polycondensed in situ either with succinyl chloride (in the case of β-BL) or with suberoyl chloride (for ε-CL) [80]. Cyclic poly(ε-CL)s were detected up to masses around 10 600 Dalton and cyclic poly(β-BL)s up to 17 000 Dalton. In addition to the cyclic polyesters linear chains having alcoholic OH and/or CO_2H group were included. This suggests that the chain growth is limited by cyclization and by incomplete conversion of the functional groups. Thus, the cyclization made a significant contribution to the conversion and to the limitation of the propagation [1, 81]. An incomplete conversion was one factor limiting the chain growth.

4.2 Cyclic Polyamides

Cyclic polyamides were reported to be isolated from Nylon 6 polymers in 1956 [18, 19]. Thermal polycondensation of ω-amino acid (carbon number ≥ 6) gave a cyclic and linear polymer [82]. Moreover, upon heating polyamide in the presence of a transamidation catalyst, the cyclization equilibrium is eventually reached, and both linear and cyclic constituents are present [83]. The proportion of the latter depends on the concentration, and cyclic compounds predominate in high dilute solutions.

4.3 Cyclic Polymers Containing Ether Linkage

Poly(ether ketone)s were prepared by polycondensation of bisphenol-A with 4,4′-difluorobenzophenone or with 2,6-difluorobenzophenone and 4′-*t*-butyl-2,6-difluorobenzophenone. Polycondensations of the free bisphenol-A was performed in DMSO or sulfolane with azeotropic distillation of water. In addition, polycondensations of bistrimethylsilyl bisphenol-A was conducted in *N*-methylpyrrolidone. The latter approach gave higher yields and higher molecular weights. The fraction of cyclic oligomers and polymers increased with an increase of the molecular weights [84]. The efficient preparation of a range of cyclic (aryl ether ketone)s containing the 1,2-dibenzoylbenzene moiety via the nucleophilic aromatic substitution route with the use of the pseudo high dilution principle was developed. Thermoanalyses show that most of these oligomers exhibit a high degree of crystallinity, while their corresponding polymers are amorphous [85].

Macrocyclic aryl ether ketone oligomers containing the 1,2-dibenzoyl-3,4,5,6-tetraphenylbenzene moiety were also synthesized in high yield by reaction of 1,2-bis(4-fluorobenzoyl)-3,4,5,6-tetraphenylbenzene with bisphenols in DMF in the presence of potassium carbonate (Fig. 23). The diketone monomer has low solubility in DMF, and high dilution conditions were obtained by adding the solid reactants in portions during the course of the reaction. The cyclic ether ketone oligomers were also transformed into cyclic ether phthalazines by reaction with hydrazine [86].

Recently, monodisperse oligo(ethylene glycol)s were polycondensed with trimesoyl chloride using triethylamine as HCl acceptor as shown in Fig. 24. Despite a nearly quantitative conversion and relatively high concentration (0.4 mol/L), no gelation occurred with tri- and tetra(ethylene glycol), regardless of the feed ratio. At concentrations of 0.2 or 0.4 mol/L di-, penta-, and hexa(ethylene glycol)s yielded gels. At concentrations of 0.2 mol/L ex-

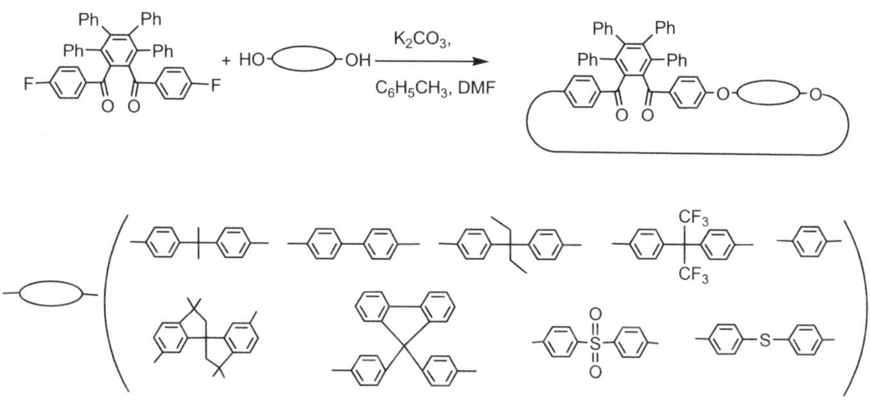

Fig. 23 Synthesis of cyclic ketone ethers containing the tetraphenylbenzene moiety

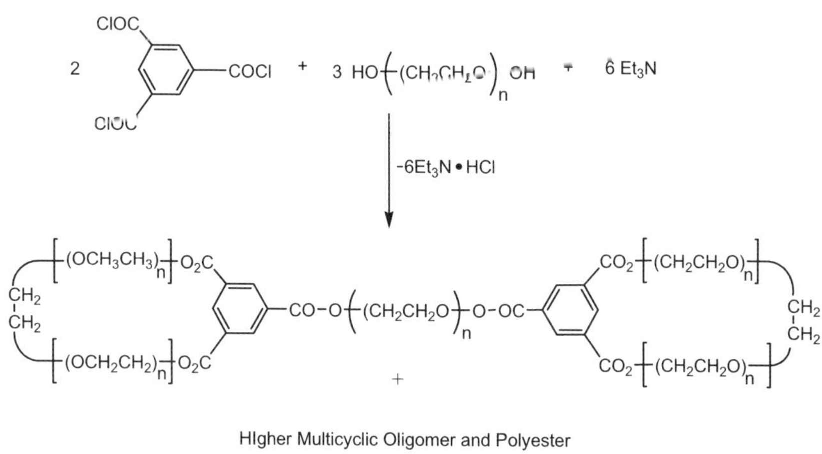

Fig. 24 Sytheses of multicyclic poly(ether ester)s from trimesoyl chloride and monodisperse oligo(ethylene glycol)s

clusively, gels were obtained from various α,ω-dihydroxyalkanes. The multicycles derived from tri- and tetraethylene glycol showed cryptant properties and dissolved Li^+, Na^+, in tetrahydrofuran [87].

4.4 Cyclic Polycarbonates

Poly(bisphenol-A-carbonate) under pseudoideal reaction conditions was investigated, and the cyclic polycarbonate was obtained as the main product. In the system, the interface of the water/toluene mixture might have favored the cyclization reaction between the polar end groups [88]. Cyclic carbonates during the (Salen)CrCl catalyzed CO_2/cyclohexene oxide copolymerization process in the presence of ionic initiators was also obtained [89]. The cyclic carbonate is produced via the backbiting mechanism, and the process is assumed to take place via a metal alkoxide (polymer chain) intermediate. Subsequent ring-opening of the cyclic carbonate with concomitant formation of polyether and CO_2 was fast at the reaction temperatures from 80 to 100 °C).

5 Synthesis of Cyclic Polymers Using Polyaddition

A reactive macrocyclic ether-urethane carbodiimide (MC-CDI), which is the key intermediate for synthesizing poly(amide-urethane) and poly(amide-imide-urethane), was prepared by the sequential self-repetitive reaction. A se-

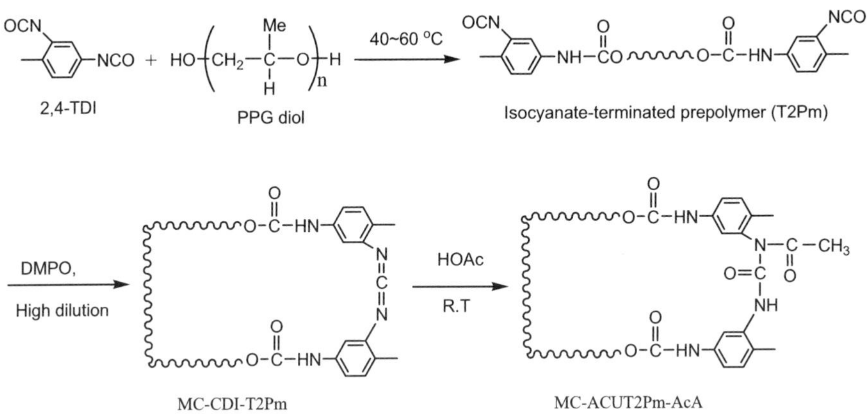

(DMPO: 1,3-dimethyl-3-phospholene oxide)

Fig. 25 Synthesis of MC-CDI and MC-ACU

ries of new aromatic MC-CDIs was synthesized and isolated the pure MC-CDIs possessing only one CDI linkage as indicated in Fig. 25.

The model reaction of the purified MC-CDI-T2Pm with acetic acid produces a cyclic acylurea (MC-ACUT2Pm-AcA), which can undergo a ring-opening reaction at above 120 °C with formation of a linear amide-isocyanate intermediate. These macrocyclic intermediates could undergo self-ring-opening, and the subsequent one-step polymerization in solution at around 120 °C formed modified polyurethanes of high molecular weight. The macrocyclic acylureas are a new latent isocyanate system and can be activated under mild conditions. This ring system seems ideal for devising a one-component urethane system or self-curing systems without the presence of the toxic isocyanate group in the formulation [90].

6
Synthesis of Cyclomers by Oxidation Polymerizations

The first cyclic aromatic disulfide, reported 40 years ago [91], was synthesized with an unspecified yield from oxidation of 1,4-naphthalenedithiol with alkaline ferricyanide and only a few other cyclic aromatic disulfides. Wong and Marvel synthesized a cyclic trimer in 30% yield by oxidation of 1,4-dimercaptobenzene with iodine in ethanol [92]. In 1991, Wang and Hay reported that poly(p-phenylene sulfide) can be prepared by heating 4,4′-diiodophenyl disulfide [93, 94]. Since then, a series of cyclic aromatic disulfide oligomers were synthesized in high yields by catalytic oxidization of arenedithiols in DMAc [95, 96]. The formation of the poly(arylene sulfide) oligomers and polymers from cyclic disulfide is depicted in Fig. 26 [97].

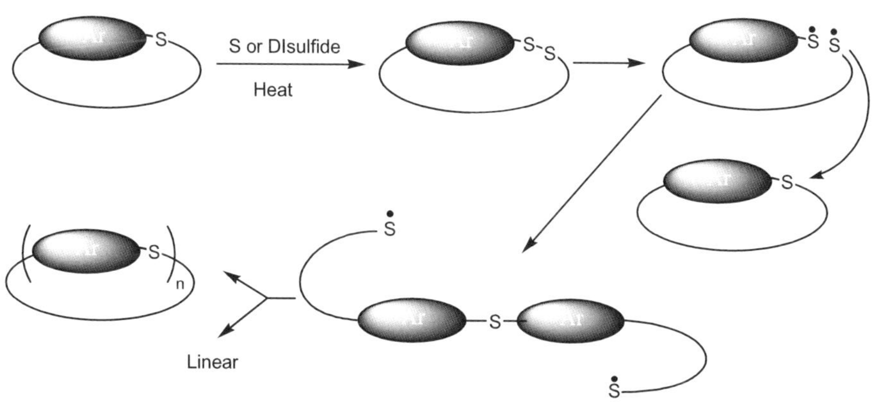

Fig. 26 Formation of the poly(arylene sulfide) oligomers and polymers

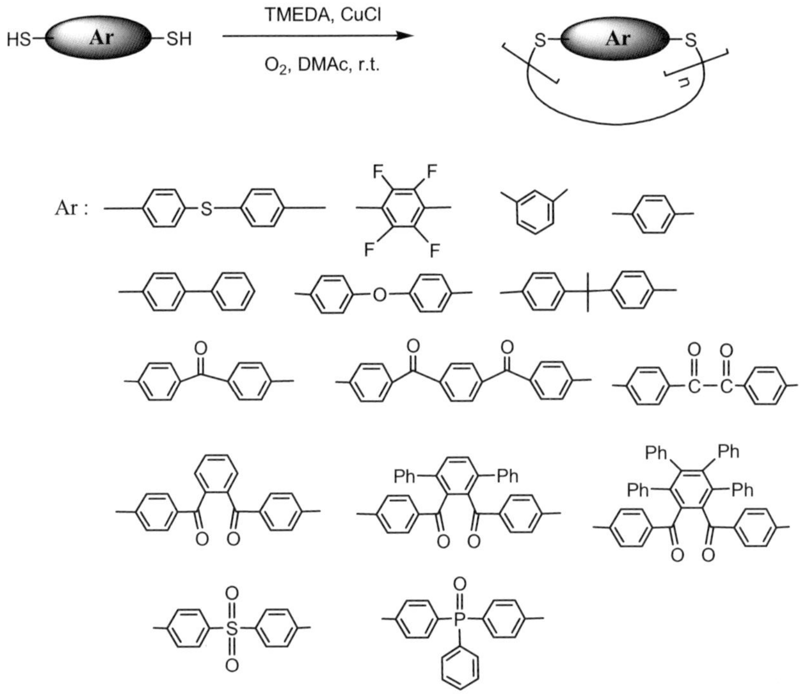

Fig. 27 Synthesis of cyclic aromatic disulfide oligomers by copper catalytic oxidation reaction

In general, the cyclic aromatic disulfide oligomers are soluble in common organic solvents and have sharp melting points. Cyclic disulfide ketone oligomers and cyclic disulfide sulfone oligomers have higher glass transition temperatures and melting points than their ether or thioether counterparts. Figure 27 shows the synthesis and characterization of a series of cyclic aromatic disulfide oligomers by copper-catalyzed oxidation of dithiol compounds with oxygen [98].

A convenient and efficient synthetic route to a new class of macrocyclic aryl ether ether sulfide oligomers was reported. The process is shown in Fig. 28. This new class of cyclic oligomers is prepared in excellent yield by quantitative chemical reduction of macrocyclic aryl ether ether sulfoxide oligomers with oxalyl chloride and tetrabutylammonium iodide. The cyclic sulfoxide oligomeric precursors are prepared in high yields by an aromatic nucleophilic substitution reaction from bis(4-fluorophenyl) sulfoxide with potassium salts of bisphenols under high-dilution conditions [99].

In the synthesis of a series of cyclic aromatic disulfide oligomers using oxidative coupling of dithiols with oxygen catalyzed by copper salts and an amine, an effective, easy, and rapid method for the synthesis of macrocyclic aromatic disulfide oligomer from 4,4′-oxybis(benzenethiol) by cyclodepoly-

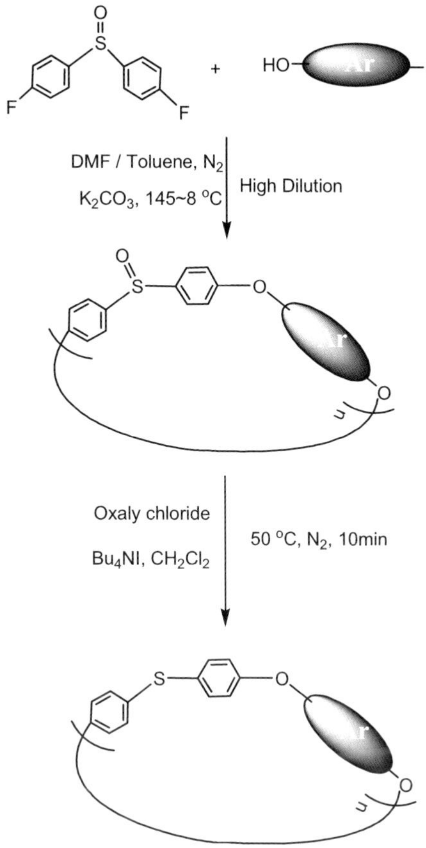

Fig. 28 Synthesis of macrocyclic aryl ether ether sulfide oligomers

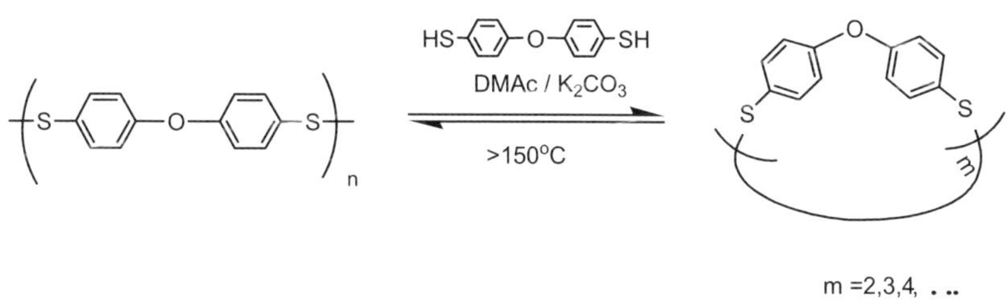

Fig. 29 Cyclopolymerization of an aromatic disulfide polymer

merization (CDP) of the linear aromatic disulfide polymer was achieved. The synthesis can be conveniently carried out on a large scale without the necessity of high-dilution conditions. The realization of the CDP reaction is demonstrated in Fig. 29. The yields were high, and the macrocyclic oligomers have a good solubility in both DMAc and THF and melt upon heating, implying the formation of low molecular weight oligomers. The cyclodepoly-

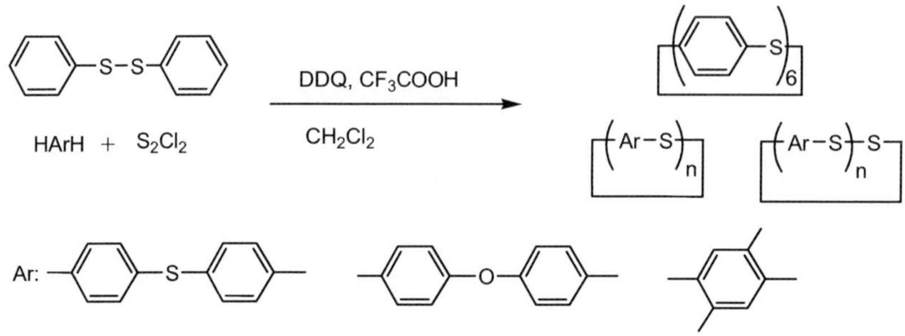

Fig. 30 Synthesis of cyclic oligo(thio arylene)s

merization of aromatic disulfide polymers is an efficient and easier way to synthesize the corresponding macrocyclic disulfide oligomers as compared with the conventional synthesis of cyclic aromatic disulfide oligomer. The novelty and merits of this work arise from (a) elimination of the need for a highly diluted condition and vigorous stirring, (b) high yield and narrow molecular weight distribution for as-made macrocyclics, (c) less costly and easier workup for purifying product, and (d) a convenient technology to recycle aromatic disulfide polymers [100].

Oxidative polymerization of aryl disulfides has been carried out under high dilution conditions for the preparation of cyclic arylene sulfides for synthesizing cyclic arylene sulfide polymers. The readily available diphenyl disulfide and dichloro disulfide with diphenyl sulfide, diphenyl ether, or *p*-xylene are useful monomers for the synthesis of the corresponding cyclic oligo(thio arylene)s. The reaction scheme is shown in Fig. 30 [101].

In addition to the above, a variety of macrocyclic aromatic compounds, including cyclic esters [102], ethers [103], amides [104, 105], ether ketones [106], and ether ether ketones [107, 108] have also been synthesized extensively in the past decade.

7
Synthesis of Cyclic Polymers by Metathesis Polymerization

Ring-opening metathesis polymerization is believed to produce cyclic polymers, but Chauvin et al. demonstrated that the polymerization of cyclopentene and cyclooctene with $WOCl_4/Et_2AlCl$ or $WOCl_4/Bu_4Sn$ catalyst afforded a linear polymer with alkylidene fragment at the chain end [109]. A new synthetic route to cyclic polymers has been developed by Grubbs et al., in which the ends of growing polymer chains remain attached to a metal complex throughout the entire polymerization process. For this purpose, catalyst $L(PCy_3)Cl_2RudCHPh$ (L = 1,3-dimesitylimidazolylidene) in conjunction with the ring-opening metathesis polymerization (ROMP) of strained cyclic

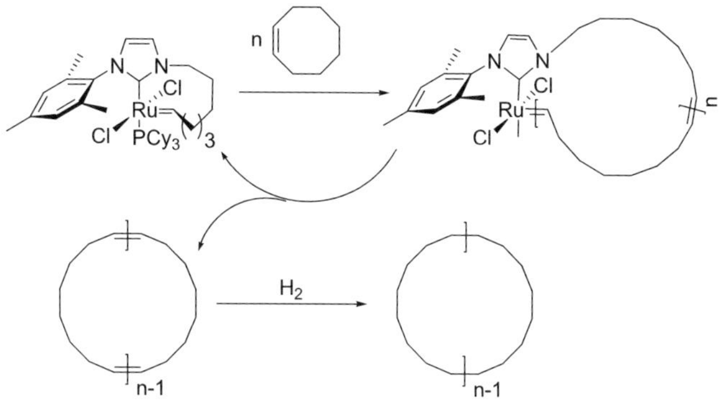

Fig. 31 Synthesis of cyclic PE using ROMP of *cis*-cyclooctene and hydrogenation

olefins was used [110]. The polymerization process is shown in Fig. 31. Polymer formation is believed to proceed through the formation of a transient macrocyclic complex in which both ends of the growing polymer chain remain attached to the Ru center. Subsequent intramolecular chain transfer later releases cyclic polymer from the intermediate complex. The high affinity of *N*-heterocyclic carbenes for Ru prevents the overall circular nature of the system from being compromised during the course of the polymerization. The approach eliminates the need for linear polymeric precursors and high dilution, drawbacks of traditional macrocyclization strategies, and it effectively removes the barrier to producing large quantities of pure cyclic material. Hydrogenation of the isolated cyclic polyoctenamers afforded the corresponding cyclic polyethylenes.

The formed cyclic polymer can be converted to linear polymer by Jones's reagent (CrO_3/H_2SO_4) as shown in Fig. 32. Polyoctenamers containing on average only one 1,2-diol group per polymer chain were obtained by adding a small amount of 1,2-diol-5-cyclooctene during the ROMP of *cis*-cyclooctene. The cyclic and linear polymers were separately reacted with Jones's Reagent that selectively cleaved the 1,2-diol unit into the corresponding bis-carboxylic acids.

The ROMP method was extended to synthesize cyclic polybutadiene using 1,5-cyclooctadiene (COD) or 1,5,9-*trans-cis-trans*-cyclododecatriene (CDT)

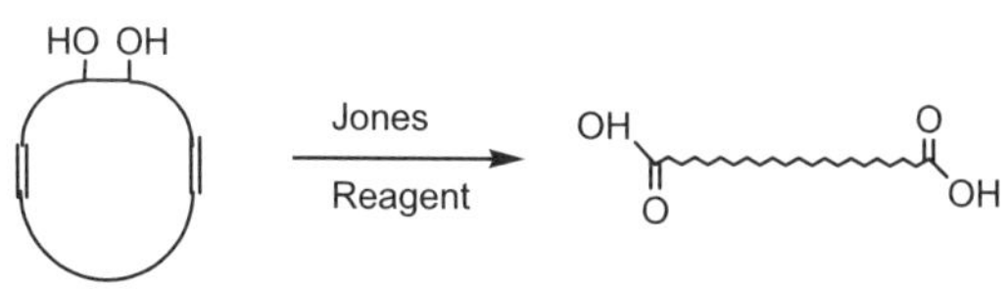

Fig. 32 Hydrogenation of the cyclic polymer containing only one 1,2-diol unit with Jones reagent

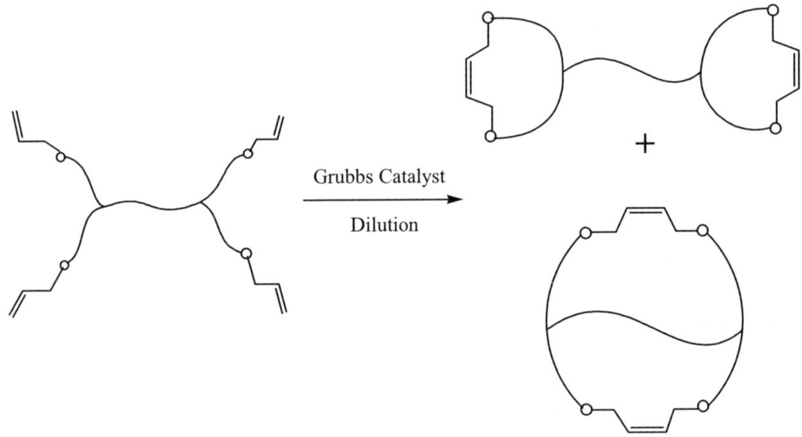

Fig. 33 Synthesis of cyclic polymer by ring closure methasesis

as the monomer. The advantage of this method affords an exclusive 1,4-regioisomeric backbone. In ROMP, the presence of an acyclic impurity, 4-vinylcyclohexene, found in the COD led to linear polymer contaminants. In contrast, CDT, which was free of the impurity, afforded cyclic polymer with little to no contamination over a wide range of molecular weights [111].

Alternatively, ring closure metathesis reaction is applied to the formation of cyclic polymers, although the main chain does not consist of hydrocarbon. Such a method was reported in the preparation of cyclic poly(THF) as shown in Fig. 33 [112]. Such ring-close metathesis was also applied for a preparation of catenane [113].

8 Synthesis of Cyclic Polymers by Ring-Opening Polymerizations

8.1 Cyclic Polyethers from Cyclic Monomers

Crown ethers, i.e., uniform cyclic oligooxyethylenes such as dibenzo-18-crown-6 was discovered by Pedersen in 1962. After his discovery, many kinds of crown ethers such as substituted polyether, sulfur- and nitrogen-containing cyclic polyethers and cryptands were synthesized [114]. The typical feature of crown ethers is able to form the metal salt complexes. However, the interest in polymer chemistry is in cyclic polymers with chain lengths long enough to generate characteristic "polymer" properties, i.e., 100–200 chain atoms or more. Such large cyclic ethers with high molecular weights were synthesized by α,ω-difunctional polyoxyethylene obtained from anionic polymerization. Ishizu et al. [115] reported that cyclic polyoxyethylenes with

$$-OCH_2CH_2OH + CH_2Cl_2 \longrightarrow -OCH_2CH_2OCH_2Cl + HCl$$

$$-OCH_2CH_2OCH_2Cl + HOCH_2CH_2O- \longrightarrow -OCH_2CH_2OCH_2-OCH_2CH_2O- + HCl$$

Fig. 34 Key reaction for synthesis of cyclic PEO

large molecular weights were prepared by the Williamson etherification reaction. Polyethylene glycol disodium salts (I) ($M_n = 8000$ and 20 000) were reacted with a stoichiometric amount of 1,4-dibromobutane in DMF, which is a similar procedure for preparation of cyclic polystyrene [58]. No intermolecular condensation took place even at high concentrations, and the conversion to cyclic polymer was substantially 100%. A preparation was described, which affords good yields of cyclic polyoxyethylenes with narrow chain length distributions and number-average molar masses in the range 1000–20 000 g mol^{-1}. The reaction was carried out in the presence of NaOH, and the process is indicated in Fig. 34 [116].

The chain folding in the crystalline state of a high molecular weight cyclic polyoxyethylene was reported [117]. Two types of photoreactive water-soluble oligo(ethylene glycol)s (OEGs; $M_n =$ ca. 6500) were prepared by derivatization of OEG with photodimerizable groups such as cinnamates or coumarinates at both ends. Coumarinate groups attached to OEG were photodimerized by UV irradiation. The molecular design of nanostructure cyclic polymers of various sizes is possible by changing the concentration of aqueous solutions [118].

Cyclic polyether is also formed by the ring-chain equilibrium in the ring opening polymerization of less strained cyclic monomers such as 1,3-dioxolane [119–121]. 1,3-Dioxacycloalkanes were polymerized in two stages, forming cyclic oligomers in the first stage and mainly high polymers in the second stage, with boron trifluoride ether complex as an initiator. All the oligomers formed in the polymerization of 1,3-dioxacycloalkanes were proved to be cycles [122].

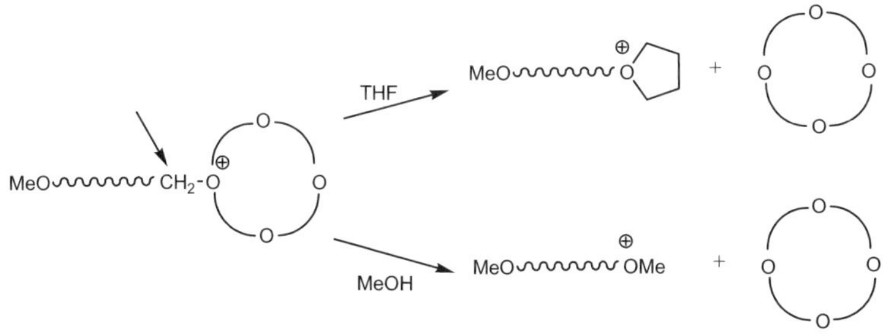

Fig. 35 Formation of cyclic polymers for the synthesis of cyclic poly(THF)

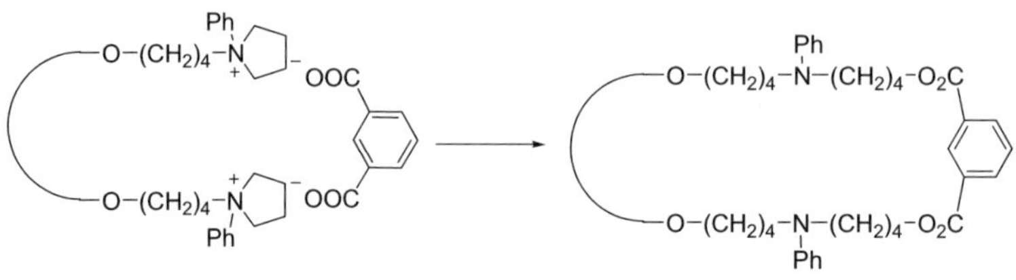

Fig. 36 Electrostatic self-assembly and covalent fixation

Concerning a cyclic polymer of tetrahtydrofuran (THF), macrocyclic oligomers of THF were reported to be synthesized with Me_3OBF_4 in CH_3NO_2 (the reaction is depicted in Fig. 35) [123].

Alternatively, Tezuka et al. have proposed an "electrostatic self-assembly and covalent fixation" process for the effective synthesis of monocyclic and three dicyclic polymer structures [124, 125]. The key reaction for formation of cyclic polymers is an electrostatic self-assembly and covalent fixation process, as indicated in Fig. 36.

They developed such a process [126–129], in which temporary stable, but specifically programmed noncovalent structures are constructed through electrostatic interaction with polymer precursors containing ionic groups and are subsequently transformed into robust and stable products of totally covalent linkages. In this process, they have purposely employed telechelic precursors having moderately strained cyclic ammonium, such as five-membered cyclic (pyrrolidinium) or six-membered bicyclic (quinuclidinium), salt groups at the designated position of a polymer chain [130]. The subsequent covalent conversion process has relied on the ring-opening reaction of the cyclic ammonium salt group by appropriately nucleophilic counteranions, such as a carboxylate, occurring at an elevated temperature. In particular, by using telechelics, having five-membered *N*-phenylpyrrolidinium salt groups, a highly controlled ring-opening reaction proceeds to form covalent aminoester groups [131], and the electrostatic self-assembly and covalent fixation process has been successfully applied for the synthesis of a variety of topologically unique polymers that have cyclic and multicyclic constructions (Fig. 37) [132–134].

A cyclic macromonomer designed for a novel polymer network architecture having both covalent and physical linkages. The methacrylate-functionalized cyclic poly(THF) has been synthesized and copolymerized through a free-radical mechanism involving methyl methacrylate and using AIBN as the initiator in benzene at 65 °C. A copolymer product with cyclic poly(THF) branches was produced initially, with gelation taking place at a later stage of the reaction. This gelation may be explained by the fact that the propagating polymer segment threaded the large but constrained cyclic polymer branches covalently attached to the polymer backbone to produce

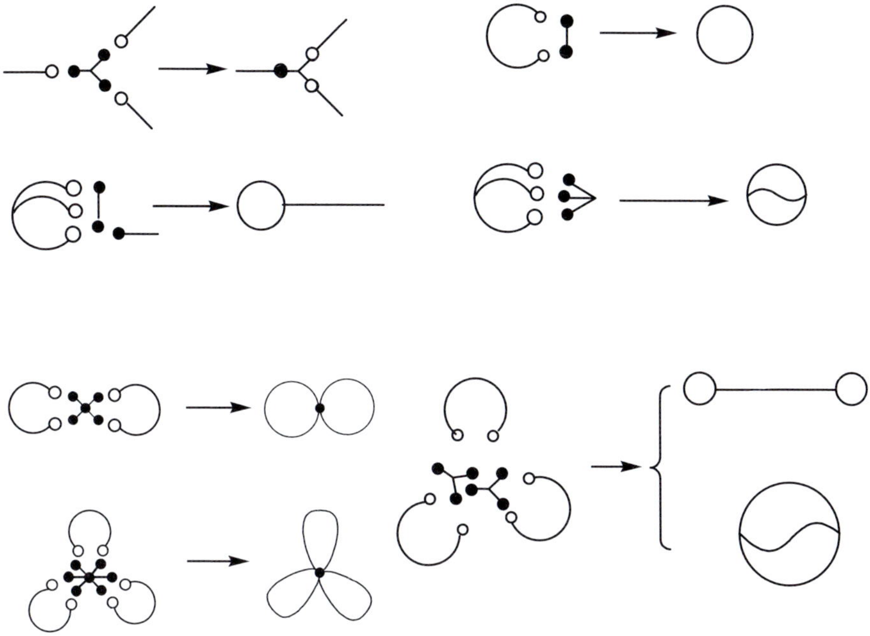

Fig. 37 Construction of multicyclic polymer topologies by electrostatic polymer self-assembly and ring-opening covalent fixation

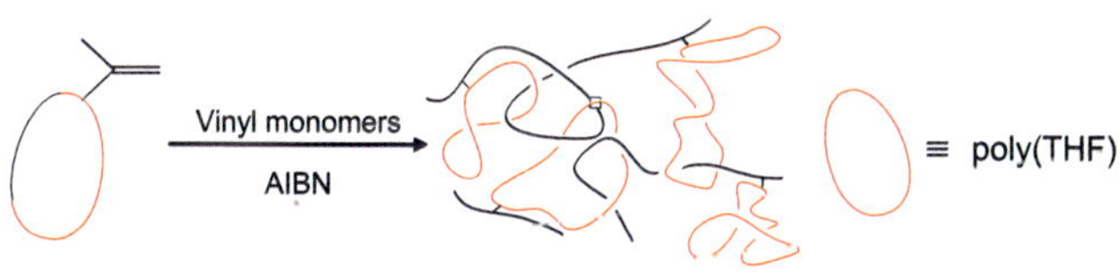

Fig. 38 Copolymerization of cyclic macromonomer of methyl methacrylate with vinyl monomers

a novel polymer network architecture, exhibiting both covalent and physical linkages as shown in Fig. 38 [135].

8.2
Cyclic Polyester from Cyclic Monomers

8.2.1
Polymerization of Lactones and Lactides

Some cyclic oligomers were formed as by-products in the polymerization of ε-caprolactone. Yamashita et al. [136] reported that the anionic polymerization of ε-caprolactone with *t*-BuOK in THF results in a living ring–chain

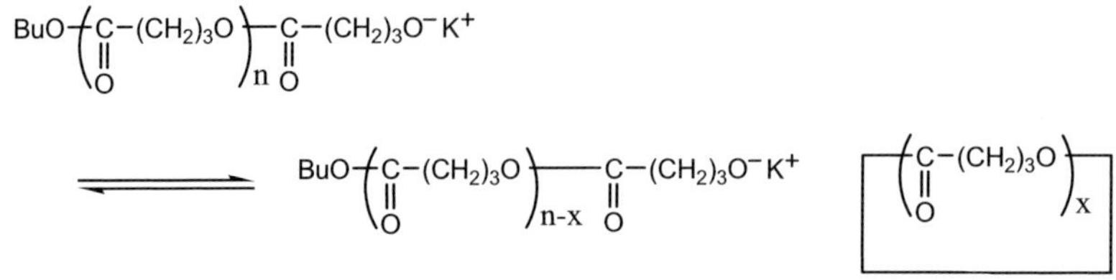

Fig. 39 Formation of cyclic oligomer in the polymerization of ε-caprolactone

equilibrium system as shown in Fig. 39. The lower cycles are favored over the linear chains at higher dilution. Terminating the reaction before establishment of the equilibrium provides direct evidence that the cyclic oligomers are produced by backbiting degradation from the initially formed linear polymers.

Recently, ring-opening polymerization of ε-caprolactone initiated by $Ti(O{-}n{-}Pr)_4$ and $Ti(OPh)_4$ was investigated. Polymerization was confirmed to proceed via a coordination-insertion mechanism for both initiators after analysis of end groups by ^1H and ^{13}C NMR spectroscopy. The molecular weight of the polymer still increases beyond the end of the conversion because of important transesterification reactions with end groups that lead to the formation of larger poly(ε-caprolactone) chains and to the "regeneration" of titanate phenoxide bonds as shown in Fig. 40. However, the intramolecular transesterification was not referred to in detail [37].

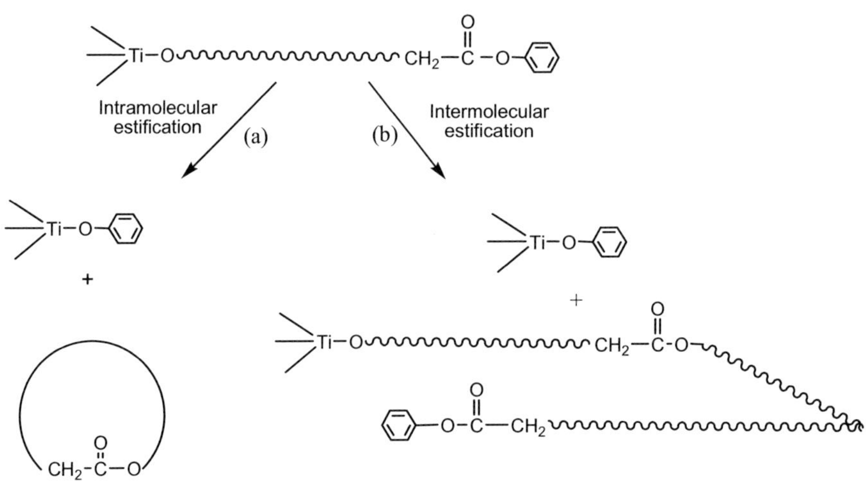

Fig. 40 Intramolecular (**a**) and intermolecular (**b**) transesterification reactions with phenoxyl ester end groups of poly(ε-caprolactone)

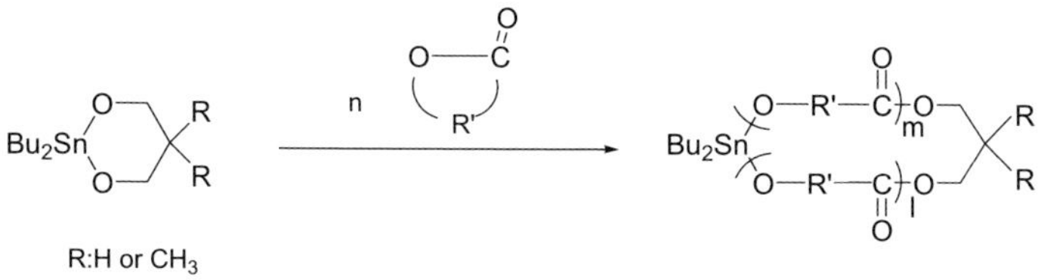

Fig. 41 Polymerization of lactone with cyclic catalyst

Kricheldorf et al. reported an anionic polymerization of γ-D,L-butyrolactone or D,L-lactide with cyclic dibutyltin initiators, such as 2,2-dibutyl-2-stanna-1,3-dioxepane, to give cyclic polymers [138–140]. Figure 41 shows the ring expansion polymerization of lactone for synthesizing a cyclic polymer as an example. They also synthesized the cyclic polymer with a living mechanism in the polymerization of ε-caprolactone [141].

This type of ring expansion polymerization was extended for a synthesis of multiblock copoly(ether-ester)s of poly(THF) and ε-caprolactone with cyclic dibutyltin initiators [142]. Such ring expansion polymerizations of lactones to synthesize macrocyclics with spirocyclic tin initiators are shown in Fig. 42 [143].

Three cyclic initiators from dibutyltin oxide and 1,2-ethanediol, 2-mercaptoethanol, and 1,2-dimercaptoethane as indicated in Fig. 43 were prepared. Polymerizations of L-lactide and racemic D,L-lactide were conducted

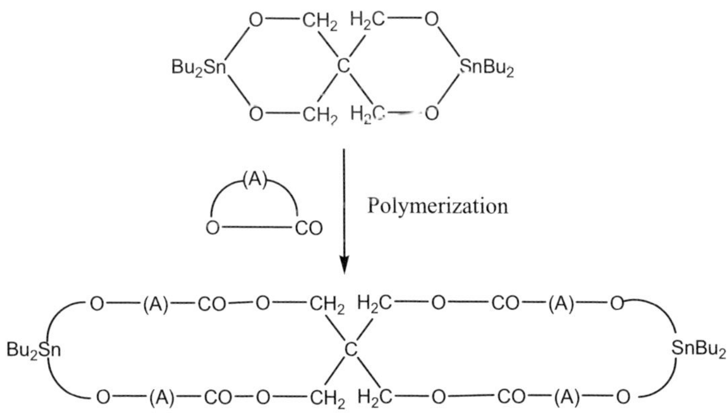

Fig. 42 Ring expansion polymerization of lactone with spiro-ring initiator

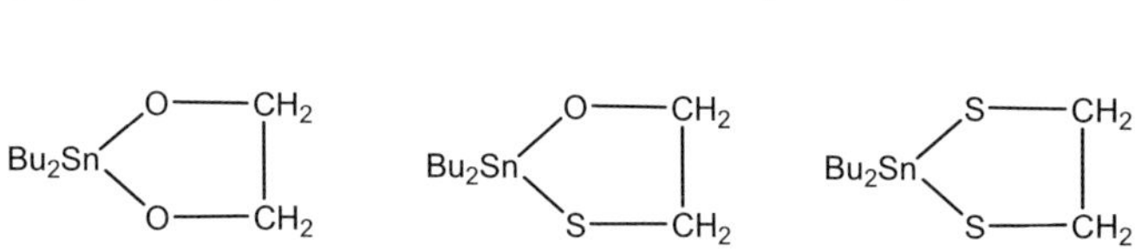

Fig. 43 Three types of cyclic tin initiators for synthesizing cyclic polyesters

in bulk at 120 °C. Among them, the 2,2-dibutyl-2-stanna-1,3-dithiolane is an extremely poor initiator. The macrocyclic polylactides resulting from the initiation with the ethanediol-based stannadioxolane can be cleaved with 1,2-dimercaptoethane, so that the dibutyltin group was selectively removed [144]. Stannylenated glucose and glycosides as cyclic initiators for the polymerization of ε-caprolactone were investigated to make biodegradable networks [145].

Star-shaped polylactones with functional end groups via ring-expansion polymerization of ε-caprolactone or γ-D,L-butyrolactone with a spiroinitiator was synthesized as shown in Fig. 44 [146]. The spirocyclic polylactones formed were reacted with various carboxylic acid chlorides and yielded four-armed stars with the elimination of Bu_2SnCl_2. Star arms with chloroacetate, 4-bromobenzoate, 4-nitrobenzoate, cinnamate, stearate, or methacrylate end groups were obtained by varying the acid chlorides.

Many works on the synthesis of cyclic polymers and block copolymers using kinetically controlled ring-expansion polymerizations of cyclic monomers, such as lactones and lactides with various types of cyclic tin initiators, were reviewed by Kricheldorf [147, 148]. Kricheldorf's group continued the synthesis of cyclic polymers, and their recent works have focused on the following. Polycondensations of 4,4′-difluorodiphenylsulfone with tris(4-hydroxy phenyl)ethane were performed in DMSO to give multicyclic poly(ether sulfone)s derived from tris(4-hydroxyphenyl)ethane [149].

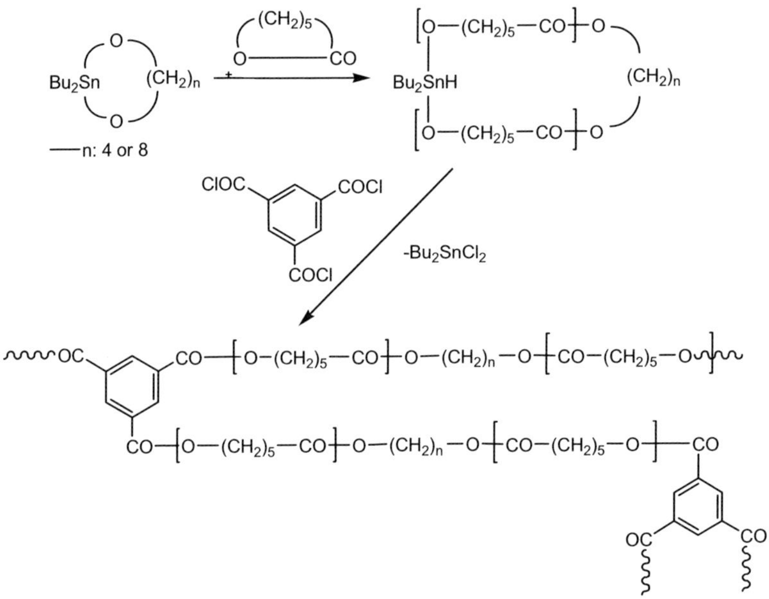

Fig. 44 Synthesis of star polymer by two stage polymerization

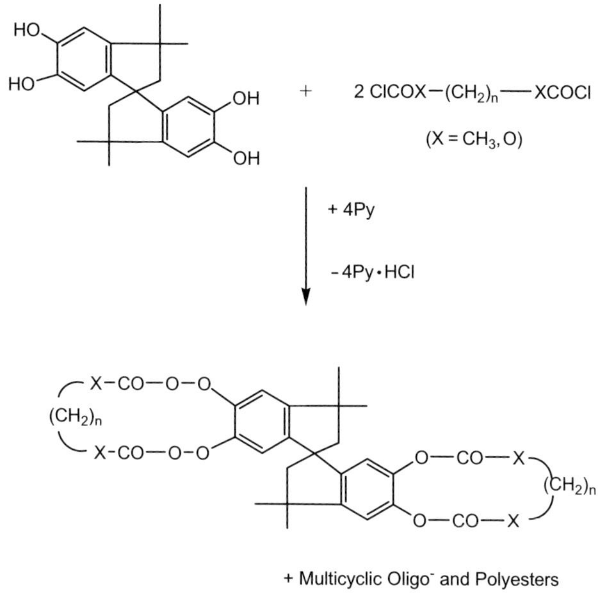

Fig. 45 Synthesis of multicyclic polyesters by polycondensation

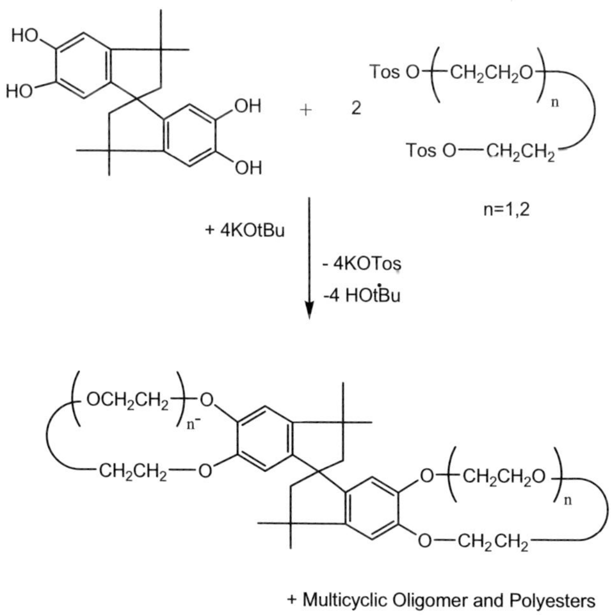

Fig. 46 Syntheses of multicyclic polyethers from TTSBI and oligo(ethylene glycol) tosylates

Multicyclic polyesters by polycondensation of 5,5′,6,6′-tetrahydroxy-3,3,3′,3′-tetramethylspiro-bisindane with sebacoyl chloride or 1,6-hexanediol bischloroformiate were synthesized, as depicted in Fig. 45, and the formed polymer has an amorphous character [150].

5,5′-Tetrahydroxy-3,3,3′,3′-tetramethyl spirobisindane (TTSBI) was deprotonated with *t*-BuOK and polycondensed with di(ethylene glycol) tosylate or tri(ethylene glycol) tosylate as shown in Fig. 46. At concentrations of 0.1 or 0.2 mol/L no cross-linking was observed and multicyclic polyethers were isolated in high yields (up to 99%). The formed polymer was an amorphous character with two glass-transition temperatures [151].

8.2.2
Lipase-Catalyzed Polymerization of Cyclic Monomers

Conventional ring-opening polymerization of cyclic anhydrides, carbonates, lactones, and lactides require extremely pure monomers and anhydrous conditions as well as metallic catalysts, which must be completely removed before use, particularly for medical applications. To avoid these difficult restrictions, an enzymatic polymerization may be one of the more feasible methods to obtain the polyesters. This method was first reported by two independent groups (Kobayashi [152] and Gutman [153]) who showed that lipases, enzymes capable of catalyzing the hydrolysis of fatty acid esters, can polymerize various medium-sized lactones.

Enzyme-catalyzed polymerization of β-lactones should offer a feasible way to create the design and synthesis of the relatively low molecular weight Poly(3HB) (HB: hydroxybutylate) and its copolymers, because lipases can accept a wide range of substrates, including cyclic lactones, to produce a wide range of polyesters. The polymerization of the (*R*)-butyrolactone (BL) was induced by lipase, and the resulting poly(3HB) containing a significant amount of cyclic polymers were produced as shown in Fig. 47. The cyclic polymer was increased with increasing monomer conversion. The (*R*)-BL was more easily polymerized by lipase to form the corresponding poly(3HB) with a higher molecular weight than that formed by using (*R*,*S*)-BL [154].

Moreover, lipase-catalyzed reactions of linear and cyclic poly(3-hydroxybutanoates) were subjected to hydrolysis, transesterification, and intramolecular esterification. A cyclic polymer along with linear polymers was pro-

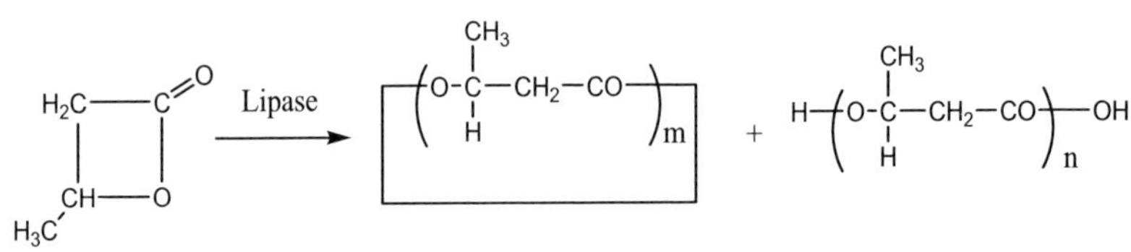

Fig. 47 Polymerization of β-butyrolactone with lipase

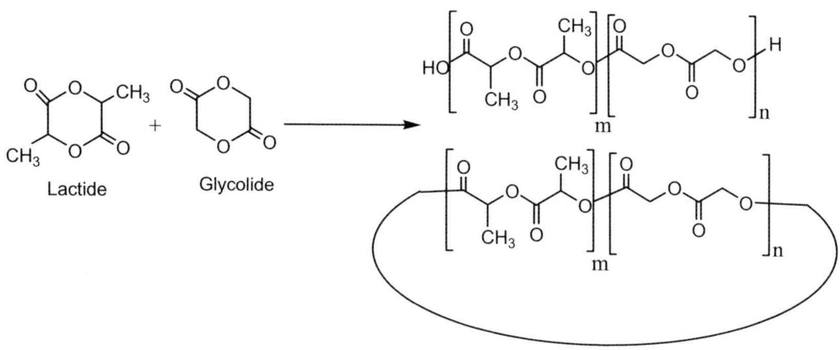

Fig. 48 Formation of poly(lactide-co-glycolide) by copolymerization of lactide and glycolide

duced by the lipase-catalyzed reaction [155]. Lipase-catalyzed copolymerization of the monomers lactide and glycolide by *Pseudomonas cepacia* was also reported. From the analysis of MALDI-TOF mass, DSC, SEC and NMR, this lipase-catalyzed copolymerization gave a linear homopolymer of lactide as the main product when the reaction carried out at 100 °C, while at 130 °C the main product was a cyclic random copolymer as depicted in Fig. 48 [156].

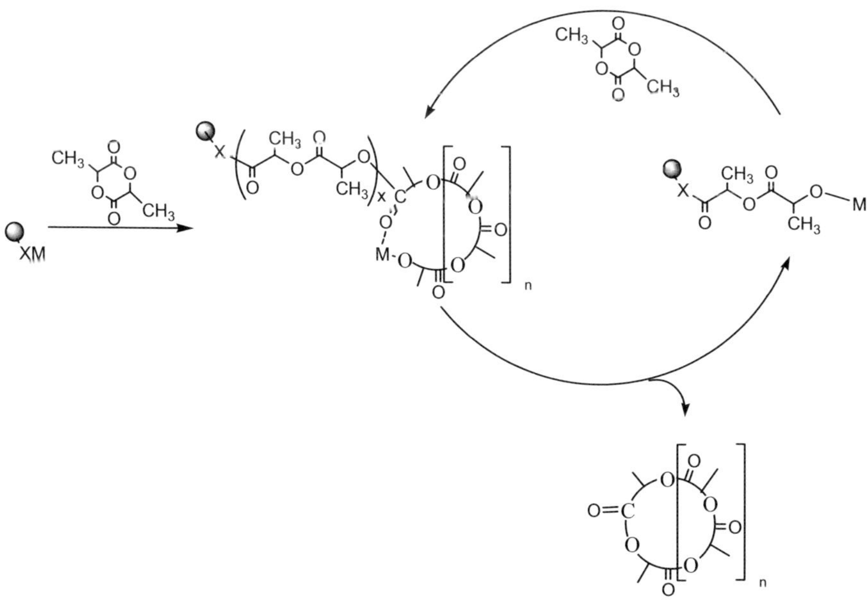

Fig. 49 Releasing of cyclic oligolactides from solid supported catalyst

Brunelle et al. [157] developed high-yielding syntheses of cyclic esters derived from 1,4-aromatic acyl chlorides and diols. Hodge et al. [158] used a solid support and step growth polymerization to produce a polyester chain followed by treatment with a metal catalyst, most often Bun_2SnO, to affect intrachain transesterification and release of the cyclic esters into the solution phase.

Bachmann and Seebach [159] have reported the preparation and characterization of cyclic lactones $(MeCHCH_2C(O)O)_n$, where $n = 4$ and 8. The reaction product between butyl lithium in benzene and the solid polystyrene support $PS\text{-}C_6H_4CH_2NH_2$ leads to a lithiated species that can be represented as $PS\text{-}C_6H_4CH_2NHLi(BuLi)_x$, where $x \sim 4$ is active in the ring-opening of the cyclic esters L-lactide, *rac*-lactide, and 2,5-morpholinediones, leading to their respective cyclic oligoesters and cyclodepsipeptides (Fig. 49) [160]. The

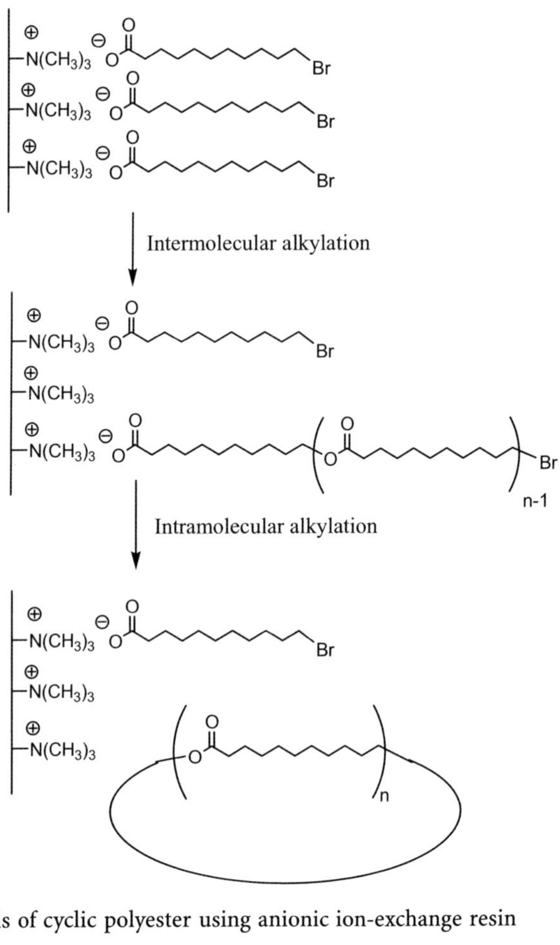

Fig. 50 Synthesis of cyclic polyester using anionic ion-exchange resin

ring enlargement occurs by the combined operations of ring opening and monomer enchainment, followed by backbiting and ring expulsion from the supported catalyst to the solution. In contrast to the conventional preparation of cyclics, high-dilution techniques are not required and no linear chains are present in the solution. The rings presented in solution can back-react with the support catalyst, so that in time the rings present will be in equilibrium and as such form a dynamic combinatorial library.

A similar solid state method for the preparation of cyclic polymer has been reported. This method involves the synthesis of cyclic polymer obtained by a combination of inter-alkylation and intra-alkylation, i.e., cyclization of 11-bromo-decanoic acid. The cyclic polymer can be liberated from an anionic exchange resin. Thus, this method has the advantage that the cyclic product from the mixture after reaction is easily produced, as indicated in Fig. 50 [161].

Cyclic polymers synthesized from biodegradable polymer such as polyglycolic acid would have unique properties with respect to the degradation rate. For example, initial breaks in the polymer chain would just convert the cyclic polymer to a straight chain polymer without affecting the molecular weight. Such a characteristic may have interesting applications to tissue engineering and drug delivery.

8.3 Cyclic Polyamide from ε-Caprolactam

From an anionic polymerization of ε-caprolactam, the following conclusion was demonstrated: The product situated between the high polymer trace and the cyclic oligomer fraction is made of macrocycles only, usually ranging from the nonamer up to M_{25} as functions of the experimental conditions of anionic synthesis. Its constituents are not at all extracted from poly(ε-caprolactam) by the usual extraction techniques, and their overall content is in the range of a few percent (2–3 wt. %), i.e., a much higher content than the values attributed to the higher oligomer fraction. MALDI-TOF mass analysis together with other characterization techniques have revealed that the amount of macrocycles made of multiples of ε-caprolactam repetition unit, typically from M_9 to M_{19}, was dependent on the experimental conditions of the synthesis, usually ca. 2.5 wt. % [162].

8.4 Cyclic Polymers from Dimethylsiloxanes

The synthesis of cyclic polydimethylsiloxane was first achieved through ring-chain equilibration of siloxane oligomers in the presence of potassium silanolate, as shown in Fig. 51 [163–165]. Cyclics recovered from ring-chain equilibration reactions have been fractionated by preparative GPC, yielding

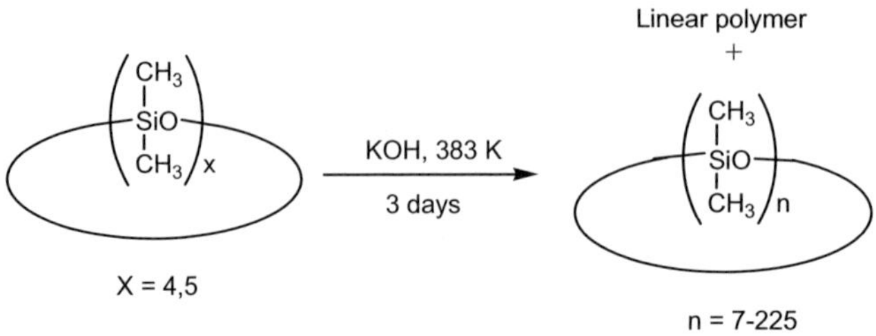

Fig. 51 Synthesis of cyclic polydimethylsiloxane by ring–chain equilibration beginning with cyclic siloxanes D4 and D5

fractions with polydispersity less than 1.05 and containing on average up to 1000 skeletal bonds. Some of the properties of these synthetic cyclic polymers were found to be in excellent agreement with theoretical predictions (for example, diluted solution viscosities), while others were not (for example, glass transition temperatures). The cyclic nature of some of the higher fractions has been demonstrated by entrapping them in network structures. Many works on the preparation, characterization and investigation of cyclic polydimethylsiloxane were performed by Semlyen et al. [10, 166–169].

Studies of cyclic and linear polydimethylsiloxanes on preparative GPC were performed [170]. Topological trapping of cyclic polymers into unimodal and bimodal model network structures was investigated [171]. The lower critical threshold temperatures (LCTT) of cyclic and linear dimethylsiloxanes, measured in both tetramethylsilane and 2,2-dimethylpropane (neopentane), was investigated. The LCTT phenomena occurred in the mixtures only if the difference in size of the polymer and solvent was sufficiently large. In all cases the cyclic polymers showed LCTT, up to 2.5 °C, by comparison with those of the linear molecules having the same number of skeletal bonds [172]. The adsorbed amounts of cyclic and linear polydimethylsiloxane fractions of narrow molecular weight distribution on silica were investigated by FTIR. At low molar masses the adsorption of cyclic polymer was greater than that of the corresponding linear polymer, but at high molar masses the reverse behavior is observed. Adsorption is greater from hexane than from CCl_4 for both linear and cyclic polymers [173].

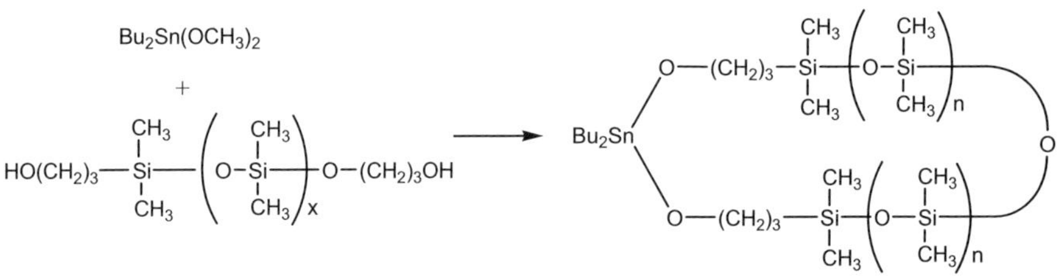

Fig. 52 Synthesis of cyclic polydimethylsiloxane by polycondensation

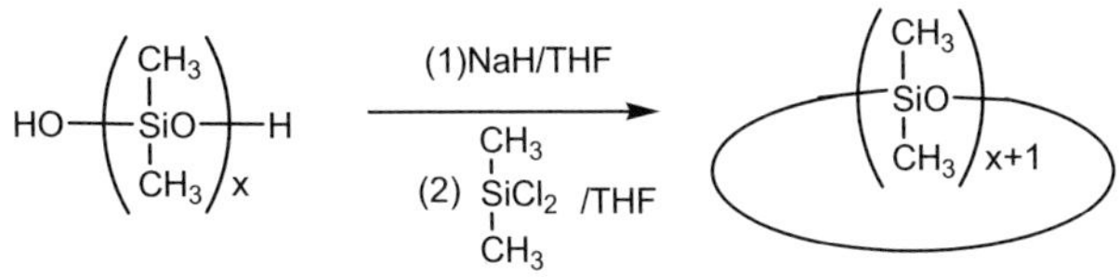

Fig. 53 Synthesis of cyclic polydimethylsiloxane for cyclization of linear α,ω-dihydroxy-polydimethylsiloxane

Surface pressure was measured as a function of surface concentration for monolayers of linear and cyclic polydimethylsiloxanes. In addition, the comparisons of linear and cyclic polymer above the overlap concentration c^* lead to the surprising conclusion that even for three-dimensional semi-dilution solutions, the ratio c/c^* was not a universal reduced concentration [174].

Cyclic siloxane-containing polymer was synthesized by the reaction of $Bu_2Sn(OCH_3)_2$ with dimethyl hydroxy-terminated polysiloxane as shown in Fig. 52 [175].

A cyclic polydimethylsiloxane was also prepared by the end-to-end reaction, i.e., α,ω-dianion-functionalized polymers, which are then cyclized simply by reaction with a difunctional electrophile to give a cyclic polymer. The cyclic polydimethylsiloxane was synthesized from a commercially available α,ω-dihydroxy-polydimethylsiloxane ($M_n = 2.460$ g/mol). The linear precursor was deprotonated using sodium hydride in dilute THF ($\leq 10^{-2}$ M) and then end-coupled using a dichlorosilane coupling agent (Fig. 53). The uncyclized anionic linear precursors are then removed by a macroporous anion exchange resin. The successful cyclization and purification is monitored by IR and ^{29}Si NMR, GPC, and MALDI-TOF mass [176].

8.5
Cyclic Polymers from Sulfur Containing Monomers

Sulfur is an interesting atom because of the specific characteristics of the simple substance and compounds containing sulfur. Further, their functions are indispensable for current technologies. Studies on sulfur compounds are valuable not only in chemistry but also in biochemistry as well. Many cyclic monomers containing sulfur such as cyclic disulfides, thiiranes, thietanes, and arene dithiols were reported to form cyclic polymers by the ring-opening polymerization. The natural source of cyclic polymers is believed to be liquid, elemental sulfur [177–179].

8.5.1 Ring-Opening Polymerization of Thioester Monomer

Cyclic polymer formation by the polymerization of thiolactone-stepwise ring-expansion and cyclic polymer contraction has been reported [180]. Recently, cyclic poly(thioglycolide) and poly(D,L-thiolactide) by a zwitter-ionic polymerization of dthiolane-2,4-diones were synthesized, as indicated in Fig. 54. Dithiolane-2,4-dione was polymerized either by heating in bulk at 140 °C or by reacting at 20 °C in the presence of pyridine or triethylamine as a catalyst. The thermal polymerization almost exclusively yielded cyclic poly(thioglycolide), but the samples prepared at 20 °C by means of tertiary amines contained cyclic and linear chains. D,L-5-methyldithiolane-2,4-dione proved so stable that tertiary amines did not catalyze its polymerization even at 100 °C. Thermal polymerization of D,L-5-methyldithiolane-2,4-dione at 160 °C yielded oligoester mainly consisting of cylics [181]. The formation of cyclic polymers is typical for the step-growth character of this polymerization process. However, it should be emphasized that zwitterionic polymerizations do not necessarily yield cyclic polymers [182].

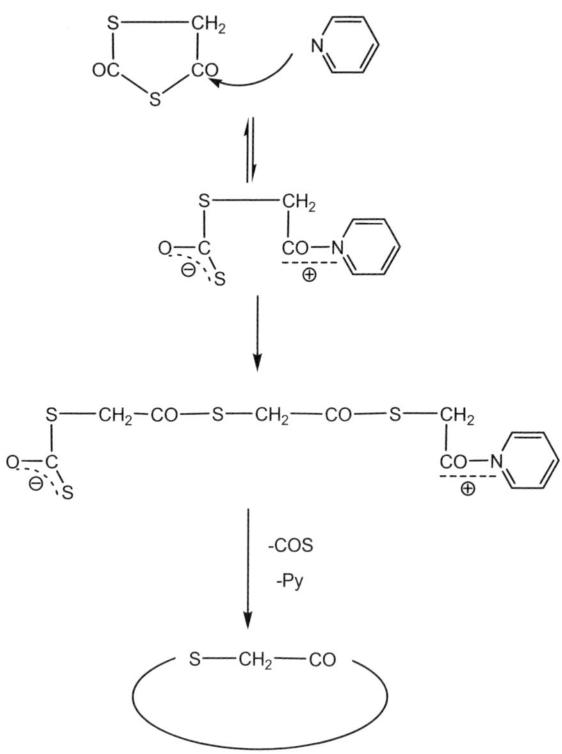

Fig. 54 Pyridine catalyzed ring-expansion polymerization of dithiolane-2,4-dione

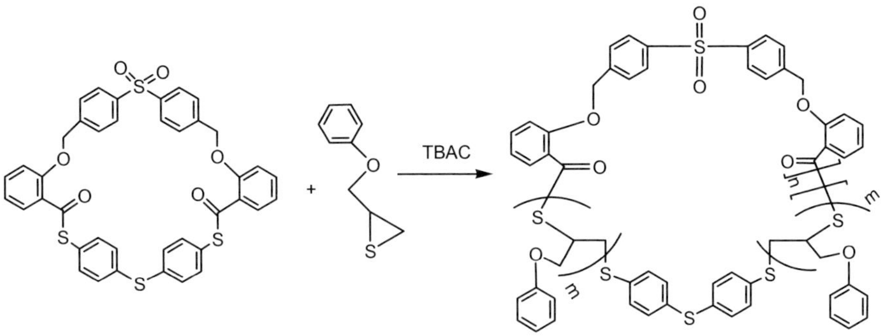

Fig. 55 Synthesis of the cyclicpolymer obtained by the insertion reaction of PPS into cyclic dithioester

The ring-expansion reaction of equimolar cyclic dithioester with 3-phenoxypropyrene sulfide (PPS) (cyclic dithioester/PPS = 1/2) was also performed in the presence of tetrabutylammonium bromide (TBAC) as a catalyst in *N*-methyl-2-pyrrolidone [183]. Intermolecular ester-exchange reaction occurred during the insertion reaction of PPS into cyclic dithioesters to give cyclic polymers with $M_n = 3600–8900$ in high yields. The structures of the cyclic polymers were confirmed by the NMR, IR, and MALDI TOF-MS spectroscopy. The continuous insertion reaction of excess PPS into cyclic dithioester proceeded quantitatively to give the corresponding cyclic polysulfide with $M_n = 42\,000$ and $M_w/M_n = 4.50$ in a 96% yield, as shown in Fig. 55. The ring-crossover polymerization of cyclic dithioester was also performed in the presence of quaternary onium salts as catalysts at 70–150 °C for 24 h in NMP. The cyclic polymers under thermodynamic control gave a lower-molecular-weight cyclic polymer with a lower polydispersity ratio [184].

8.5.2 Ring-Opening Polymerization of Thiaethers

A novel method for synthesizing polythiaether macrocycles with Re cluster complex was proposed by Adams and coworkers [185]. This process involves catalysis by the Re complex rather than the stoichiometric displacement of dithiol or dithiolates by organic halides, and first example of a catalytic procedure for synthesis of polythiaether macrocycle using thietanes. The macrocyclic products formed are shown in Fig. 56.

$Re_2(CO)_9(NCMe)$ also activates β-propiothiolactone toward the opening of the heterocyclic ring under mild conditions to give polythiolactone macrocycles, as shown Fig. 57 [186]. However, a half amount of product was linear polymer in the reaction.

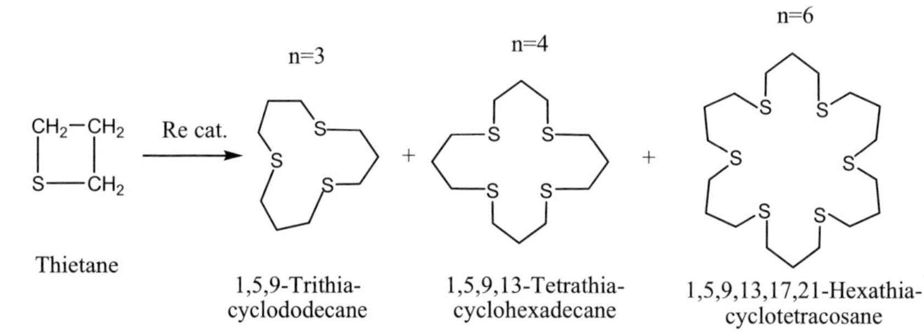

Fig. 56 Re cluster catalyzes macrocyclic formation

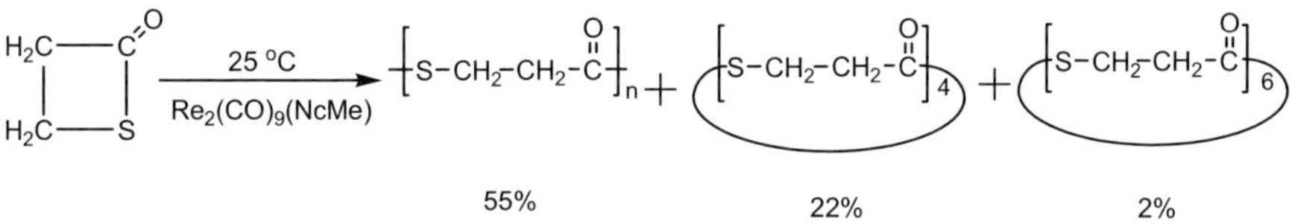

Fig. 57 Synthesis of cyclic oligomerization of β-propiothiolactone

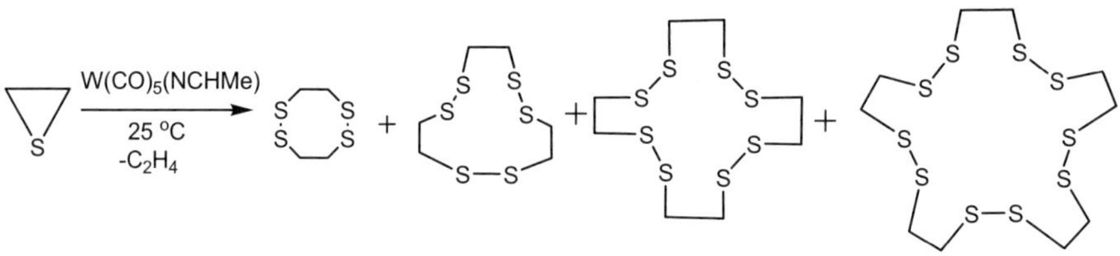

Fig. 58 Synthesis of the cyclic polydisulfides with $W(CO)_5$-(NCMe)

Moreover, the reactions of an excess of thiirane with $W(CO)_5(NCMe)$ catalyst led to the formation of the cyclic polydisulfides as depicted in Fig. 58 [187].

8.5.3 Ring-Opening Polymerization of Cyclic Disulfides

The studies on the polymers containing disulfide linkages in the main chain have been investigated [188, 189], and the biggest factor to control the reaction may be the specific behavior, that is, the cleavage of the formed disulfide linkage by heat, photo, or electrochemical reduction [190, 191]. The following are given as typical methods for synthesizing the polymers containing disulfide bonds in the main chains: oxidative coupling reactions of α,ω-dithiols [192], polycondensations of aromatic dihalides with sulfur [193], and ring-opening polymerizations of cyclic disulfide compounds.

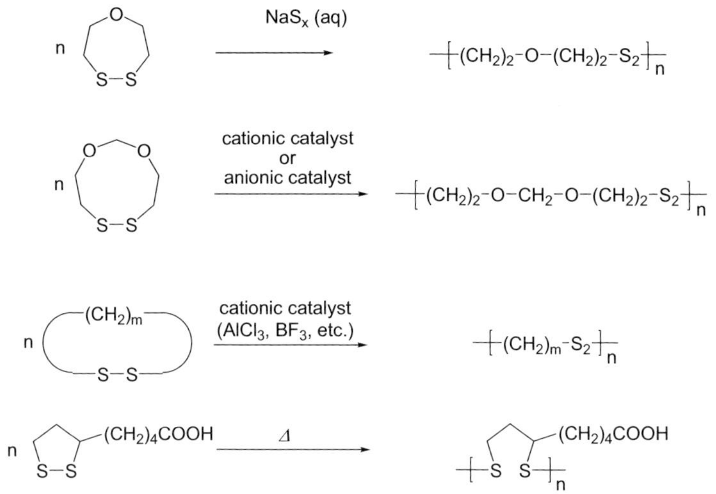

Fig. 59 Ring-opening polymerization of cylic disulfides

Among them, polymerizations of cyclic disulfides have been investigated since the 1940s (Fig. 59). Davis et al. reported that 1-oxa-4,5-dithiacycroheptane very rapidly polymerizes to rubbery polymers with aqueous sodium sulfide [194]. Tobolsky reported that the polymerization of 1-oxa-4,5-dithiacycroheptane and 1,3-dioxa-6,7-dithiacyclononane proceeds by ionic catalysts [195]. 1,2-Dithiacycloalkanes also polymerize with a trace of $AlCl_3$ or BF_3 [196]. Photo-irradiation was used for the polymerization of lipoic acid [197]. Moreover, the ring-opening polymerization of the cyclic disulfide trimer has been briefly studied to evaluate the stability of the cyclic product [198]. These reactions are expected to make cyclic polymers by intramolecular cyclization.

Although many studies on the polymerizations of cyclic disulfides have been reported, the detailed structural analyses of the resulting polymers and their characterizations have not been investigated for a long time. Endo et al. reported the thermal polymerization of cyclic disulfide (the examined monomers are shown in Fig. 60). Among the cyclic disulfides examined, the polymerization of 1,2-dithiane (DT) and the structural analyses and characterizations of the resulting polymers were investigated in detail [199]. They investigated the polymerization by use of extremely pure DT monomer (not containing any thiols or other impurities). The thermal polymerization of DT in bulk without any initiators proceeded readily and gave a polymer with a high molecular weight in a high yield when the disulfide monomer was heated above the melting point (31–32 °C). The polymerization of DT was not induced at the monomer concentration below the 4.0 mol/L in the monomer feeds. The polymerization of DT was inhibited by an addition of radical inhibitors, and the polymer yield decreased remarkably in the presence of thi-

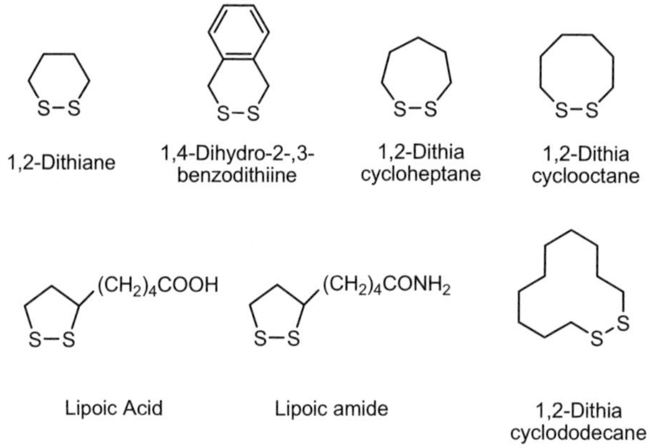

Fig. 60 Cyclic disulfides family

ols, indicating that the propagation proceeds by a radical intermediate [200]. The relationship between the DT concentration in the polymerization and the resulting polycatenane structure was also investigated [201].

From the analyses of NMR and electron-spray ionization mass (ESI-MS) spectroscopy, the polymers obtained from the polymerization of cyclic disulfides were found to be a cyclic structure [202]. The cyclic structure consisting of poly(DT) is assumed to be formed by a backbiting reaction of propagating species [203]. Thermal and mechanical properties of the polymers, and decomposition behaviors of the polymers demonstrate that the polymers obtained from thermal polymerization of cyclic disulfides include a polycatenane structure. From polymerization of cyclic disulfides in the presence of cyclic poly(ethylene oxide), a polycatenane consisting of two different cyclic polymers was obtained [199]. Thus, poly(DT) contains spatial entanglements of cyclic polymers with each other (a polycatenane structure was presumed) (Fig. 61).

Copolymerization of lipoic acid (LPA) and 1,2-dithiane (DT) was also induced readily to give high polymers in high yields, and the resulting structure contained polycatenane structures [203]. The polymers revealed characteris-

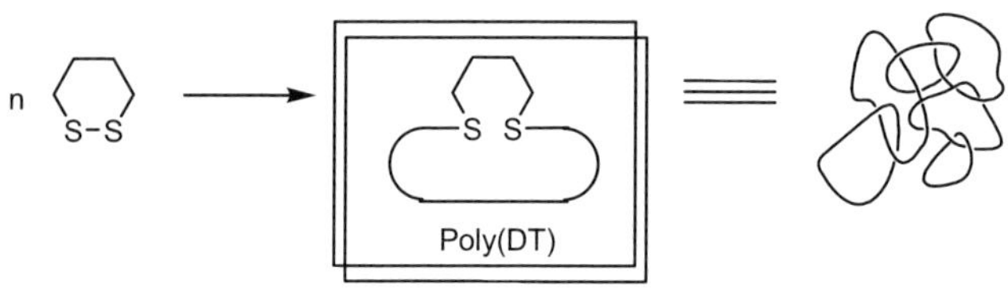

Fig. 61 Structure of poly(DT)

tics as a shape-memory material [204, 205]. Gel containing catenane structure was obtained from copolymerization [206].

The structure of polymer obtained from thermal polymerization of cyclic disulfides was suggested to include an interlocked structure. This is another consequence of the ring topology [207] Cyclic polymers not only can be branched or cross-linked, but can also form noncovalently linked structures based on their loop topology. Thus, cyclic polymers can be used as a component of rotaxanes consisting of linear chains threaded through cyclic molecules and catenane, which consist of mechanically-linked cyclic polymers. Many cyclic polymers will be used as a component for building interlocked materials. The chemistry of polyrotaxanes and polycatenanes has gained much attention, and many articles and reviews have been published by various researchers [208–225].

9 Metal Containing Cyclic Polymers

The self-assembly of a novel supramolecular cyclic polymer species of general formula $[Ru_3O(CH_3COO)_6(CO)(pz)]_6$ has been achieved by starting from μ-oxoruthenium cluster units in the presence of bridging pyrazine (pz) ligands, as shown in Fig. 62 [226].

Cyclic polymers and the block copolymers bearing alkoxy and vinylidene groups by Pd-complex-promoted living polymerization of 2-alkoxy-1-methylene-cyclopropanes were synthesized. A living ring-opening polymerization of 2-alkoxy-1-methylenecyclo-propanes was achieved by π-allyl Pd complexes to afford polymers having alkoxy and vinylidene groups in the repeating units as shown in Fig. 63. A cyclic dinuclear Pd complex was synthesized and utilized for the synthesis of cyclic polymers and block copolymers with narrow molecular weight distribution [227].

Organometallic macrocycles and cyclic polymers were prepared by the photolytic ring opening of a silicon-bridged ferrocenophane with a bipyridine initiator. The relative amounts of cyclic oligomers and cyclic polymer, as well as the molecular weight of the cyclic polymer, can be controlled by the reaction temperature [228].

10 Physical Properties of Cyclic Polymers

There have been many interesting theoretical studies of the average conformations and physicochemical behavior of cyclic polymers. The ratio of the radii of gyration of a random coil linear polymer to that of a cyclic polymer was theoretically predicted. The mean square of radius of gyration ($\langle S^2 \rangle$) of

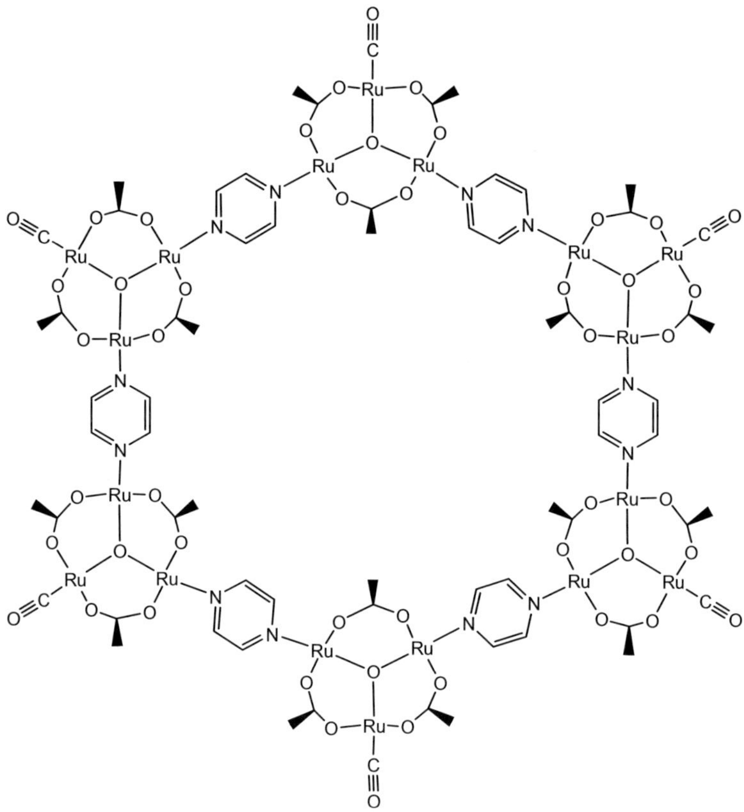

Fig. 62 Self-assembly of a novel supramolecular cyclic polymer

linear and cyclic polymers with the same total number of segments N can be calculated under the assumption of Gaussian statistics. For the linear polymer, the equation is as follows: $\langle S^2 \rangle_{\text{linear}} = Nb/6$, in which b is the length of each segment. For the cyclic polymer, the equation of $\langle S^2 \rangle_{\text{cyclic}} = Nb/12$ is conducted. In other words, in the absence of a volume effect, the ratio of $g_s \equiv \langle S^2 \rangle_{\text{cyclic}} / \langle S^2 \rangle_{\text{linear}}$ should be exactly two in the case of the same molecular weight. From the molecular dependence of the $\langle S^2 \rangle$ for cyclic polystyrene, the g_s was determined to be around 2 [229, 230]. From neutron scattering measurement of the polydimethylsiloxane ($M_w < 2.0 \times 10^4$) in benzene, the g_s was estimated to be close to 0.53, indicating that the expansion of the coil in a good solvent is somewhat higher than for the linear polymer [231].

The limiting-viscosity number ($[\eta]$) was measured at 34.5 °C for linear and cyclic polystyrene in cyclohexane, which is θ solvent for polystyrene at 34.5 °C. The relationship between $[\eta]$ and M_w is shown in Fig. 64. The ratio of $[\eta]_{\text{cyclic}}/[\eta]_{\text{linear}}$ was found to be close to 0.66, and the exponent of the vis-

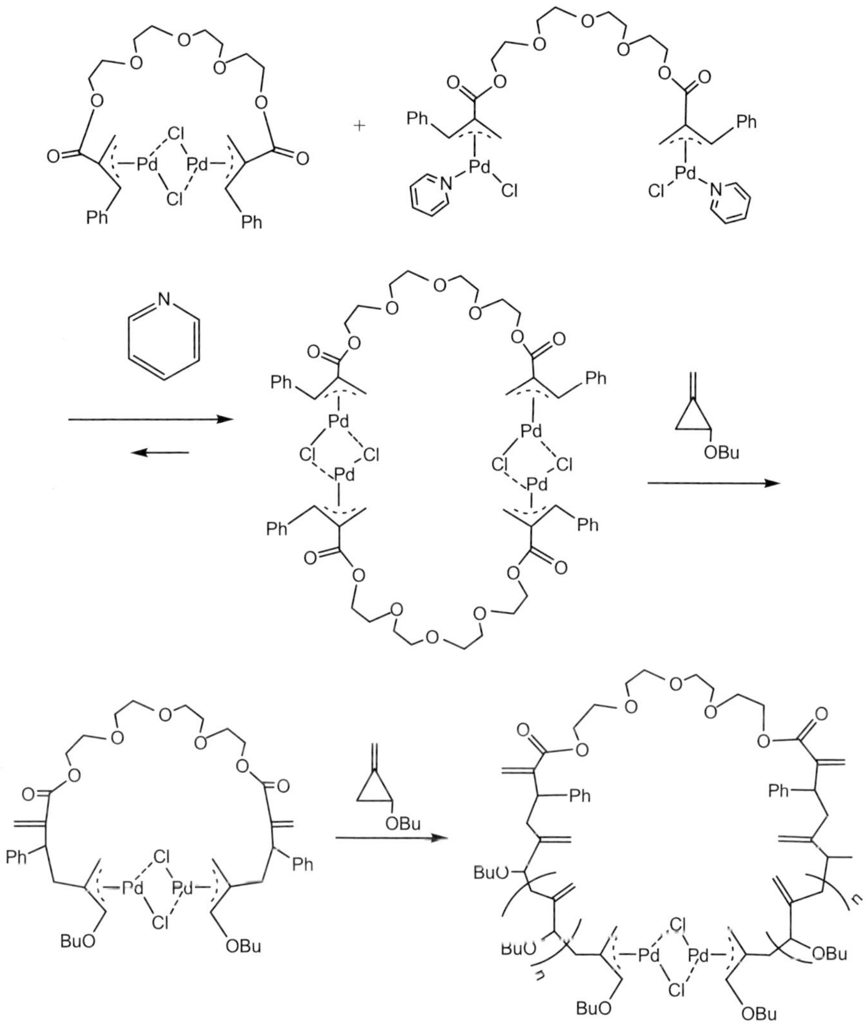

Fig. 63 Formation of cyclic polymer with controlled molecular weight

cosity law for cyclic polymers was 0.53 (in contrast to linear polymer, which was 0.50) [232].

A difference in volume between cyclic and linear polymers leads to different solution properties of the cyclic polymers compared to the linear polymers, such as: higher GPC elution volume [233], lower intrinsic viscosity [234], lower translational friction coefficient [235], neutron scattering functions; more rapid decrease of second virial coefficient with molecular weight [236], and higher critical solution temperature [237].

The cyclic polymers have also shown profound differences from linear polymer in other properties including higher density [238, 239] melt vis-

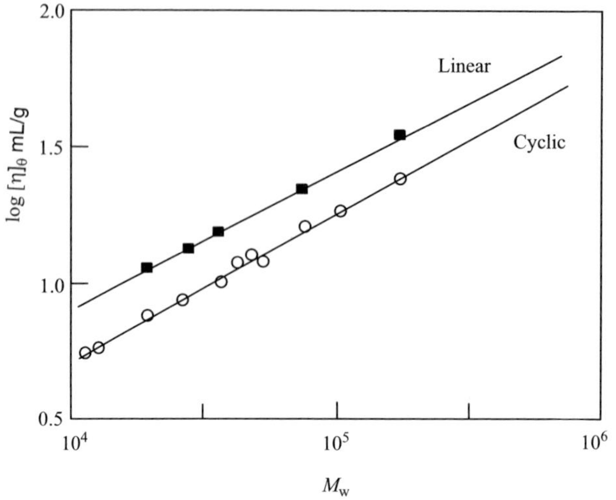

Fig. 64 Relation between $[\eta]$ and M_w for linear and cyclic polystyrene measured in cyclohexane at 34.5 °C

cosity: greater at low molecular weights, but smaller at high molecular weights [238], higher refractive index [239], lower dipole moment at low molecular weights [240], and higher glass transition temperature [241–243]. The absence of chain ends has led to speculation that the dynamic properties of polymeric rings differ from their linear analogues [244, 245].

Monte Carlo calculation of hydrodynamic properties such as the radius of gyration, friction coefficient, and intrinsic viscosity of cyclic polystyrene and polydimethylsilozane in ideal solution was investigated [246]. The size of a ring polymer in a ring-linear blend was also investigated. In melts, ring polymers assume more compact conformations compared to linear chains. These predictions were tested by performing Monte Carlo simulations of ring-linear blends using the bond-fluctuation model [247].

11 Conclusion

Cyclic polymers are now of great interest from the view point of not only their unique properties but also for the component of interlocked polymers, such as polyrotaxnenes and polycatenanes. Development of analytical instruments makes it possible to confirm detailed structures of polymers. The polymers with a well-defined structure can be synthesized using living polymerization of vinyl monomers, leading to the synthesis of a variety of cyclic polymers. The growth and commercialization of cyclic polymers has been stunted due to problems associated in synthesizing larger-cyclic polymers with routine monomers and without linear chain contamination. Despite this historic limitation, cyclic polymers will continue to have a place in polymer synthesis through further development of synthetic methods. Comprehension of the relationship of structural properties of cyclic polymers may open a new door for their application in many fields.

References

1. Flory PJ (1953) Principles of Polymer Chemistry. Cornell University Press, Ithaca New York
2. Gudipati CS, Finlay JA, Callow JA, Callow ME, Wooley KL (2005) Langmuir 21:3044
3. Lee CC, Yoshida M, Fréchet JMJ, Dy EE, Szoka FC (2005) Bioconjugate Chem 16:535
4. Chen H-T, Neerman MF, Parrish AR, Simanek EE (2004) J Am Chem Soc 126:10044
5. Harth E, van Horn B, Lee YY, Germack DS, Gonzales CP, Miller RD, Hawker CJ (2002) J Am Chem Soc 124:8650
6. Dichtel WR, Hecht S, Fréchet JM (2005) J Org Lett 7:4451
7. Thomas KRJ, Thompson AL, Sivakumar AV, Bardeen CJ, Thayumanavan S (2005) J Am Chem Soc 127:373
8. Helms B, Liang CO, Hawker CJ, Fréchet JMJ (2005) Macromolecules 38:5411
9. Garcia-Martinez JC, Lezutekong R, Crooks RM (2005) J Am Chem Soc 127:5097
10. Semlyen JA (ed) (2000) Cyclic Polymers, 2nd edn. Kluwer Academic, Dordrecht, The Netherlands
11. Rempp P, Strazielle C, Lutz P (1987) Macrocyclic Polymers. In: Kroschwitz JI (ed) Encyclopedia of Polymer Science and Engineering, Vol. 9, 2nd edn. Wiley, New York, p 183
12. Jacob F, Wollman EL (1958) Symp Soc Exp Biol 12:75–92
13. Gnanou Y (1996) J Macromol Sci Part C Polym Rev 36:77
14. Jacobson H, Stockmayer WHJ (1950) Chem Phys 18:1600
15. Jacobson H, Beckmans C, Stockmayer WH (1950) J Chem Phys 18:1607
16. Ross SD (1954) J Polym Sci 13:406
17. Goodman I, Nesbitt BF (1960) Polymer 1:384
18. Hermans PH, Heikens D, van Velden PF (1955) J Polym Sci 16:451
19. Hermans PH (1956) Nature 177:126
20. Geiser D, Hocker H (1980) Macromolecules 13:653
21. Hild G, Kohler A, Rempp P (1980) Eur Polym J 16:525
22. Vollmert B, Huang JX (1981) Makromol Chem Rapid Commun 2:467
23. Roovers J, Toporowski PM (1983) Macromolecules 16:843
24. Quirk RP, Ma J (1988) Polym Prepr 29(2):10
25. Bedha MC, Gibson HW (1990) Polym Prepr 31(1):588
26. Schappacher M, Deffieux A (1991) Makromol Chem Rapid Commun 12:447
27. Schappacher M, Deffieux A (1992) Macromolecules 25:6744
28. Deffieux A, Schappacher M, Riquet-Lurbet L (1994) Polymer 35:4562
29. Deffieux A, Beinat S, Schappacher M (1997) Macromol Symp 118:247
30. Rique-Lubet L, Schappacher M, Deffieux A (1994) Macromolecules 27:6318
31. Otsu T, Yoshida M (1982) Makromol Chem Rapid Commun 3:133
32. Kamigaito M, Ando T, Sawamoto M (2001) Chem Rev 101:3689
33. Matyjaszewski K, Xia J (2001) Chem Rev 101:2921
34. Braunecker WA, Matyjaszewski K (2007) Prog Polym Sci 32:93

35. Hawker CJ, Bosman AW, Harth E (2001) Chem Rev 101:3661
36. Moad G, Rizzardo E, Thang SH (2005) Aust J Chem 58:379
37. Bénédicte L, Xavier P, Michéle M, Patrick H (2001) Macromolecules 34:425
38. Tao H, Gen-Hua Z, Cai-yuan P (2003) Macromolecules 36:5960
39. Hua D, Ge X, Tang J, Zhu X, Bai R, Pan C (2007) J Polym Sci Part A Polym Chem 45:2847
40. Kolb HC, Finn MG, Sharpless KB (2004) Angew Chem Int Ed 2001:40
41. Huisgen R (1984) 1,3-Dipolar Cycloaddition Chemistry. Wiley, New York
42. Boyd AL, Scott MG (2006) J Am Chem Soc 128:4238
43. Qiu X, Tanaka F, Winnik FM (2007) Macromolecules 40:7069
44. Xu J, Ye J, Liu S (2007) Macromolecules 40:9103
45. Whittaker MR, Goh Y-K, Gemici H, Legge TM, Perrier S, Monteiro MJ (2006) Macromolecules 39:9028
46. Szwarc M, Levy M, Milkovich R (1956) J Am Chem Soc 78:2656
47. Casassa EF (1965) J Polym Sci A 3:605
48. Geisert D, Höcker H (1980) Macromolecules 13:653
49. Hild G, Kohler A, Rempp P (1980) Eur Polym J 16:525
50. Rempp P, Lutz P (1992) Makromol Chem Macromol Symp 62:213
51. Vollmert B, Huang J (1981) Makromol Chem Rapid Commun 2:467
52. Gan Y, Dong D, Hogen-Esch TE (1995) Macromolecules 28:383
53. Ohtani H, Kotsuji H, Momose H, Matsushita Y, Noda I, Tsuge S (1999) Macromolecules 32:6541
54. Lescanec RL, Hajduk DA, Kim GY, Gan Y, Yin R, Gruner SM, Hogen-Esch TE, Thomas EL (1995) Macromolecules 28:3485
55. Ohtani H, Kotsuji H, Momose H, Matsushita Y, Noda I, Tsuge S (1999) Macromolecules 32:6541
56. Benedicte L, Marie-Anne D, Michele M, Patrick H, Katia B, Henri C (2000) Macromolecules 33:8218
57. Boutillier J-M, Lepoittevin B, Favier J-C, Masure M, Hémery P, Sigwalt P (2002) Eur Polym J 38:E243–E250
58. Ishizu K, Kanno H (1996) Polymer 37:1487
59. Iatrou H, Hadjichristidis N, Meier G, Frielinghaus H, Monkenbusch M (2002) Macromolecules 35:5426
60. Cho D, Masuoka K, Koguchi K, Asari T, Takano A, Matsushita Y (2005) Polym J 37:506
61. Takano A, Kushida Y, Aoki K, Masuoka K, Hayashida K, Cho D, Kawaguchi D, Matsushita Y (2007) Macromolecules 40:679
62. Takano A, Kadoi O, Hirahara K, Kawahara S, Isono Y, Suzuki J, Matsishita Y (2003) Macromolecules 36:3045
63. Pasch H, Deffieux A, Henze I, Schappacher M, Rique-Lurbet L (1996) Macromolecules 29:8776
64. Pasch H, Deffieux A, Ghahary R, Schapacher M, Rique-Lurbet L (1997) Macromolecules 30:98
65. Roovers J, Toporowski PM (1983) Macromolecules 16:843
66. Roovers J, Toporowski PM (1988) J Polym Sci Part B Polym Phys 26:1251
67. Kubo M, Hayashi T, Kobayashi H, Tsuboi K, Itoh T (1997) Macromolecules 30:2805
68. Kubo M, Hibino T, Tamura M, Uno T, Itoh T (2002) Macromolecules 35:5816
69. Kubo M, Kato N, Uno T, Itoh T (2004) Macromolecules 37:2762
70. Mizawa T, Takenaka K, Shiomi T (2000) Polym Sci Part A Polym Chem 38:237
71. Kubo M, Yamamoto H, Uno T, Itoh T, Sato H (2001) Polym Bull 47:25
72. Kubo M, Nishigawa T, Uno T, Itoh T, Sato H (2003) Macromolecules 36:9264
73. Lee KS, Wegner G (1985) Makromol Rapid Commun 6:203

74. Lee KS, Wegner G, Hsu SL (1987) Polymer 28:889
75. Repin H, Papanikolan E (1969) J Polym Sci A-1 7:3426
76. Montaudo G (1991) Macromolecules 24:5829
77. Kricheldorf HR, Langanke D, Spickermann J, Schmidt M (1999) Macromolecules 32:3559
78. Kricheldorf HR, Lorenc A, Spickermann J, Maskos M (1999) J Polym Sci Part A Polym Chem 37:3861
79. Kricheldorf HR, Rabentein M, Maskos M, Schmidt M (2001) Macromolecules 34:713
80. HR Kricheldorf, D Langanke, A Stricker, Räder HJ (2002) Macromol Chem Phys 203:405
81. Mark H, Whitby GS (eds) (1940) Collected Papers of WM Carothers on Polymerization. Interscience, New York
82. Yumoto H, Ogata N (1958) Makromol Chem 25:91
83. Andrewa JA, Jones FR, Semlyen JA (1974) Polymer 15:420
84. Kricheldorf HR, Garaleeh M, Schwarz G, Vakhtangishvili L (2004) High Performance Polymers. Macromolecules 16:137
85. Chan KP, Wang Y, Hay AS, Hronowski XL, Cotter RJ (1995) Macromolecules 28:6705
86. Ding Y, Hay AS (1996) Macromolecules 29:3090
87. Kricheldorf HR, Lomadze N, Polefka C, Schwarz G (2006) Macromolecules 39:2107
88. Horbac A, Vernaleken H, Weirauch K (1980) Makromol Chem 181:111
89. Donald J. D, Paolo B, Jeremy R (2007) Macromolecules 40:7727
90. Chen C, Cheng C, Dai SA (2007) Macromolecules 40:8139
91. Marschalk C (1952) Bull Soc Chim Fr 147
92. Wong DT, Marvel CS (1976) J Polym Sci 14:1637
93. Wang ZY, Hay AS (1991) Macromolecules 24:334
94. Wang ZY, Hay AS (1992) Macromol Symp 54/55:247
95. Ding Y, Hay AS (1996) Macromolecules 29:4811
96. Wang Y, Hay AS (1995) Macromolecules 28:5050
97. Wang Y, Chan KP, Hay AS (1995) Macromolecules 28:6371
98. Ding Y, Hay AS (1996) Macromolecules 29:6386
99. Wang Y-F, Hay AS (1997) Macromolecules 30:182
100. Liang ZA, Meng YZ, Li L, Du XS, Hay AS (2004) Macromolecules 37:5837
101. Tsuchida E, Miyatake K, Yamamoto K, Hay AS (1998) Macromolecules 31:6469
102. Hubbard P, Brittain WJ, Simonsick WJ Jr, Ross CW (1996) Macromolecules 29:8304
103. Ganguly S, Gibson HW (1993) Macromolecules 26:2408
104. Kim YH, Calabrese J, McEween C (1996) J Am Chem Soc 118:1545
105. Memeger W Jr, Lazar J, Ovenall D, Leach RA (1993) Macromolecules 26:3476
106. Colquhoun HM, Dudman CC, Thomas M, O'Mahoney CA, Williams DJ (1990) J Chem Soc Chem Commun p 336
107. Jonas A, Legras R (1993) Macromolecules 26:2674
108. Chen M, Gibson HW (1996) Macromolecules 29:5502
109. Herisson JL, Cauvin Y (1971) Makromol Chem 141:161
110. Bielawski CW, Benitez D, Grubbs RH (2002) Science 297:2041
111. Bielawski CW, Benitez D, Grubbs RH (2003) J Am Chem Soc 125:8424
112. Tezuka Y, Ohashi F (2005) Macromol Rapid Commun 26:608
113. Wang L, Vysotsky MO, Bogdan A, Bolte M, Böhmer V (2004) Science 304:1312
114. Pedersen CJ (1978) Introduction – history and overviews. In: Izatt RM, Christensen JJ (eds) Synthetic multidentate macocyclic compounds, Chap 1. Academic Press, New York
115. Ishizu K, Akiyama Y (1997) Polymer 38:491

116. Yu G-E, Sinnathamby P, Price C, Booth C (1996) Chem Commun p 31
117. Cooke J, Viras K, Yu G-E, Sun T, Yonemitsu T, Ryan AJ, Price C, Booth C (1998) Macromolecules 31:3030
118. Nakayama Y, Matsuda T (2005) J Polym Sci Part A Polym Chem 43:3324
119. Szymanski R, Kubisa P, Penczek S (1983) Macromolecules 16:1000
120. Rentsh C, Schulz RC (1977) Makromol Chem 178:2535
121. Chwiakouska W, Kubisa P, Penczek S (1982) Makromol Chem 183:753
122. Kawakami Y, Yamashita Y (1977) Macromolecules 10:837
123. McKenna JM, Wu TK, Pruckmayr G (1977) Macromolecules 10:877
124. Oike H, Imaizumi H, Mouri T, Yoshioka Y, Uchibori A, Tezuka Y (2000) J Am Chem Soc 122:9592
125. Tezuka Y, Oike H (2001) Macromol Rapid Commun 22:1017
126. Tezuka Y (2005) Chem Rec 5:17
127. Tezuka Y (2003) J Polym Sci Part A Polym Chem 41:2905
128. Tezuka Y, Oike H (2002) Prog Polym Sci 27:1069
129. Tezuka Y, Oike H (2001) Macromol Rapid Commun 22:1017
130. Tezuka Y (1996) Telechelic Oligomers (with Cyclic Onium Salt Groups). In: Salamone JC (ed) Polymeric Materials Encyclopedia, Vol 11. CRC, Boca Raton FL, p 8263
131. Oike H, Imamura H, Imaizumi H, Tezuka Y (1999) Macromolecules 32:4819
132. Tezuka Y, Fujiyama K (2005) J Am Chem Soc 127:6266
133. Oike H, Imaizumi H, Mouri T, Yoshioka Y, Uchibori A, Tezuka Y (2000) J Am Chem Soc 122:9592
134. Kaoru A, Takasugi H, Tezuka Y (2006) Macromolecules 39:5585
135. Oike H, Mouri T, Tezuka Y (2001) Macromolecules 34:6229
136. Ito K, Hashizuka Y, Yamashita Y (1977) Macromolecules 10:821
137. Cayuela J, Bounor-Legaré V, Cassagnau P, Michel A (2006) Macromolecules 39:1338
138. Kricheldorf HR, Lee S-R, Bush S (1996) Macromolecules 29:1375
139. Kricheldorf HR, Lee S-R (1996) Macromolecules 29:8689
140. Kricheldorf HR, Eggerstedt S (1998) Macromol Chem Phys 199:283
141. Kricheldorf HR, Eggerstedt S (1998) Macromolecules 199:283
142. Kricheldorf HR, Langanke D (1999) Macromol Chem Phys 200:1183
143. Kricheldorf HR, Rost S (2004) Polymer 45:3205
144. Kricheldorf HR, Lee S-R, Bush S (1996) Macromolecules 29:1375
145. Kricheldorf HR, Stricker A (2000) Macromolecules 33:696
146. Kricheldorf HR, Fechner B (2002) J Polym Sci Part A Polym Chem 40:1047
147. Kricheldorf HR (2004) J Polym Sci Polym Chem Ed 42:4723–4742
148. Kricheldorf HR, Chatti S, Schwarz G (2004) Macromol Symp 210:93
149. Garaleh M, Polefka C, Schwarz G, Kricheldorf HR (2007) Macromol Chem Phys 208:747
150. Kricheldorf HR, Bornhorst K, Schwarz G (2007) Macromolecules 40:199
151. Kricheldorf HR, Lomadze N, Schwarz G (2007) Macromolecules 40:4818
152. Uyama H, Kobayashi S (1993) Chem Lett 22:1149
153. Knani D, Gutman AL, Kohn DH (1993) J Polym Sci Part A Polym Chem 31:1221
154. Matsumura S, Suzuki Y, Tsukada K, Toshima K, Doi Y, Kasuya K (1998) Macromolecules 31:6444
155. Osanai Y, Toshima K, Matsumura S (2000) Chem Lett 29:576
156. Huijser S, Staal BBP, Huang J, Duchateau R, Koning CE (2006) Biomacromolecules 7:2465
157. Brunelle DJ, Bradt JE, Serth-Guzzo J, Takekoshi T, Evans T, Pearce EJ,Wilson PR (1998) Macromolecules 31:4782

158. Ruddick CL, Hodge P, Cook A, McRiner A (2002) J Chem Soc Perkin Trans 1:629
159. Bachmann BM, Seebach D (1998) Helv Chim Acta 81:2430
160. Chisholm MH, Gallucci JC, H Yin, (2006) Proc Natl Acad Sci USA 103:15315
161. Wood BR, Hodge P, Semlyen JA (1993) Polymer 34:3052
162. Casazza E, Ricco L, Russo S, Scamporrino E (2007) Macromolecules 40:739
163. Brown JF, Slusarczuk GM (1965) J Am Chem Soc 87:931
164. Dodgson K, Semlyen JA (1977) Polymer 18:1265
165. Clarson SJ, Semlyen JA (1986) Polymer 27:1633
166. Semlyen JA (1981) Pure Appl Chem 53:1797
167. Semlyen JA (1986) Cyclic Polymers. Elsevier, New York
168. Semlyen JA (1996) Large Ring Molecules. Wiley, New York
169. Semlyen JA, Wood BR, Hodge P (1994) Polym Adv Technol 5:473
170. Dodgson K, Sympson D, Semlyen JA (1978) Polymer 19:1285
171. Clarson SJ, Mark JE, Semlyen JA (1987) Polym Commun 28:151
172. Barbarin-Castillo JM, McLure IA, Clarson SJ, Semlyen JA (1987) Polym Commun 28:212
173. Patel A, Cosgrove T, Semlyen JA (1991) Polymer 32:1313
174. Granick S (1985) Macromolecules 18:1597
175. Kricheldorf HR, Langanke D (1999) Macromol Chem Phys 200:1174
176. White BM, Watson WP, Barthelme EE, Beckham HW (2002) Macromolecules 35:5345
177. Semlyen JA (1967) Trans Faraday Soc 63:2342
178. Semlyen JA (1968) Trans Faraday Soc 64:1396
179. Semlyen JA (1971) Polymer 12:383
180. Kricheldorf HR, Lee S-R, Schittenhelm N (1998) Macromol Chem Phys 199:273
181. Kricheldorf HR, Lomadze N, Schwarz G (2007) Macromolecules 40:4859
182. Kricheldorf HR, Garaleh M, Schwarz G (2005) J Macromol Sci Pure Appl Chem A42:139
183. Kudo H, Makino S, Kameyama A, Nishikubo T (2005) Macromolecules 38:5964
184. Kudo H, Makino S, Nishikubo T (2007) J Polym Sci Part A Polym Chem 45:680
185. Adams RD, Falloon SB (1994) J Am Chem Soc 116:10540
186. Adams RD, Huang M, Huang W, Queisser JA (1996) J Am Chem Soc 118:9442
187. Adams RD, Queisser JA, Yamamoto JH (1996) J Am Chem Soc 118:10674
188. Kishore K, Ganesh K (1995) Adv Polym Sci 121:81
189. Duda A, Penczek S (1989) Rotaxanes. In: Kroschwitz JL (ed) Encyclopedia of Polymer Science and Engineering, Vol 16, 2nd edn. Wiley, New York, pp 246–368
190. Nambu Y, Acar MH, Suzuki T, Endo T (1988) Makromol Chem 189:495
191. Ganesh K, Kishore K (1996) Macromolecules 29:26
192. Choi W, Sanda F, Kihara N, Endo T (1998) J Polym Sci Part A Polym Chem 36:79
193. Lowman DW, Fagerburg DR (1993) Macromolecules 26:4606
194. Davis FO, Fettes EM (1948) J Am Chem Soc 70:2611
195. Tobolsky AV, Leonard I, Roeser GP (1948) J Polym Sci 3:604
196. Dainton FS, Ivin KJ, Walmsley DAG (1960) Trans Faraday Soc 56:1784
197. Barltrop JA, Hayes PM, Calvin M (1954) J Am Chem Soc 76:4348
198. Houk J, Whitesides GM (1989) Tetrahedron 45:91
199. Endo K, Shiroi T, Murata N, Kojima G, Yamanaka T (2004) Macromolecules 37:3143
200. Endo K, Shiroi T, Murata N (2005) Polym J 37:512
201. Endo K, Kojima G, Yamanaka T (2006) Kautsch Gummi Kunstst 59:41
202. Arakawa R, Watanabe T, Fukuo T, Endo K (2000) J Polym Sci Part A Polym Chem 38:4403
203. Endo K,Yamanaka T (2006) Macromolecules 39:4038

204. Endo K (2005) Kobunnsi 54:805
205. Endo K (2007) Nippon Gomu Kyokaishi 80:191
206. Yamanaka T, Endo K (2007) Polym J 39:1360
207. Gibson HW (1996) Large Ring Molecules. John Wiley, New York
208. Gibson HW, Bheda MC, Engen PT (1994) Prog Polym Sci 19:843
209. Gong C, Gibson HM (1997) Opin Solid State Mater Sci 2:647
210. Gibson HW (1996) Rotaxanes. In: Semlyen JA (ed) Large Ring Molecules, Chap 6. John Wiley and Sons, New York, p 191
211. Raymo FM, Stoddard JF (1999) Chem Rev 99:1643
212. Gong C, Gibson HM (1999) Polyrotaxane: Synthesis and Properties. In: Sauvage J-P, Dietrich-Buchecker CO (eds) Molecular Catenanes, Rotaxanes and Knot, Chap 11. Wiley-VCH, Weinheim, p 277
213. Mahan H, Gibson HW (2000) Rotaxanes. In: Semlyen JA (ed) Cyclic Polymers, Chap 12, 2nd edn. Kluwer Publisher, Dordrecht, p 415
214. Hubin IN, Busch DH (2000) Chem Rev 200–202:5
215. Panov IG, Topchieva IN (2001) Russ Chem Rev 70:23
216. Huang F, Gibson HM (2005) Prog Polym Sci 30:982
217. Harada A, Hashidzume A, Takashima T (2006) Adv Poly Sci 201:1
218. Wasserman E (1960) J Am Chem Soc 82:4433
219. Greets Y (1999) Polycatenanes, poly[2]-catenanes and polymeric catenanes. In: Sauvage JP, Dietrich-Buchecker CO (eds) Molecular Catenanes, Rotaxanes, and Knots. Wiley-VCH, Weinheim
220. Sauvage J-P (1998) Acc Chem Res 31:611
221. Fujita M (1999) Acc Chem Res 32:53
222. Wang L, Vysotsky MO, Bogdan A, Bolte M, Böhmer V (2004) Science 304:1312
223. Chichak KS, Cantrill SJ, Pease AR, Chiu S-H, Cave GWV, Atwood JL, Stoddart JF (2004) Science 304:1308
224. Raymo FM, Stoddart JF (1999) Chem Rev 99:1643
225. Takata T (2006) Polym J 38:1
226. Toma HE, Nikolaou S (2000) J Chem Res (S) p 326
227. Takeuchi D, Inoue A, Osakada K, Kobayashi M, Yamaguchi K (2006) Organometallics 25:4062
228. Chan WY, Lough AJ, Manners I (2007) Angew Chem Int Ed 46:9069
229. Casassa EF (1965) J Polym Sci 3:605
230. Higgins JS, Dodgson K, Semlyen JA (1979) Polymer 20:553
231. Roover J (1985) J Polym Sci Polym Phys Ed 23:1117
232. Strazielle C, Lutz P, Rempp P (1986) Makromol Rapid Commun 7:599
233. Freifelder D, Kleinschmidt AK, Sinscheimer RL (1964) Science 146:254
234. Clarson SJ, Semlyen J A (1986) Polymer 27:1633
235. Edwards CJC, Stepto RFT, Semlyen JA (1980) Polymer 21:781
236. Edwards CJC, Stepto RFT, Semlyen JA (1982) Polymer 23:869
237. Barbarincastillo JM, Mclure IA, Clarson SJ, Semlyen JA (1987) Polym Commun 28:212
238. Orrah DJ, Semlyen JA, Ross-Murphy SB (1988) Polymer 29:1455
239. Bannister DJ, Semlyen JA (1981) Polymer 22:377
240. Beevers MS, Mumby SJ, Clarson SJ, Semlyen JA (1983) Polymer 24:1565
241. Clarson SJ, Semlyen JA (1986) Polymer 27:1633
242. Chien YY, Winter WT (1985) Macromolecules 18:1357
243. Gan YD, Dong DH, Hogen-Esch TE (1995) Macromolecules 28:383
244. Obhukov SP, Rubinstein M, Duke T (1994) Phys Rev Lett 73:1263

245. McLeish T (2002) Science 297:2005
246. Garcia Bernal JM, Tirado MM, Freire J, Garcia de la Torre (1991) Macromolecules 24:593
247. Iyer BVS, Ashish KL, Sachin S (2007) Macromolecules 40:5995

Subject Index